A First Course in Technical English

Teachers' Book 2

LYNETTE BEARDWOOD
HUGH TEMPLETON
MARTIN WEBBER

HEINEMANN EDUCATIONAL BOOKS
LONDON

HEINEMANN EDUCATIONAL BOOKS LTD
22 Bedford Square, London WC1B 3HH

LONDON EDINBURGH MELBOURNE AUCKLAND
HONG KONG SINGAPORE KUALA LUMPUR NEW DELHI
IBADAN NAIROBI JOHANNESBURG EXETER (NH)
KINGSTON PORT OF SPAIN

ISBN 0 435 28759 1
(Students' Book ISBN 0 435 28758 3)
(Tape ISBN 0 435 28760 5)
(Cassette ISBN 0 435 28031 7)

First published 1979
Reprinted 1980, 1981, 1982, 1984

Cover: 'Night photograph of discharges over experimental insulators at Brighton Insulator Testing Station of the Central Electricity Research Laboratories, Leatherhead.' Reproduction by permission of the Central Electricity Generating Board.

Phototypeset in Great Britain by
Filmtype Services Limited, Scarborough
Printed and bound in Great Britain by
Spottiswoode Ballantyne Ltd.,
Colchester and London

CONTENTS

Introduction

A First Course in Technical English is intended for foreign students needing English for their technical studies. Book 2 carries straight on from Book 1, providing practice and revision at an elementary/intermediate language level. The materials are the result of research and experimental teaching at a large technical college overseas where competence in English was a prerequisite for studying technical subjects.

Course Design

The course is based on a series of illustrated texts, each specially written to highlight and practise a limited number of language features. The content is not too 'technical' since the aim is to teach English in a technical context, not technology through English.
Emphasis is placed on developing an ability to comprehend spoken and written English through the texts. Attention is also given to the students' needs for writing and speaking English through the exercises.
Language items are selected for their functional value, i.e. for their relevance to students' needs in using technical language. These language items are developed methodically throughout the course with thorough oral and written practice.

Contents of Book 2

Book 2 is divided into eight *Units* which deal with different technical topics. Each Unit is divided into *Sections*, containing a Reading Text and four or five exercises. *Most Sections also have a related but separate Listening Text which only appears in the Teachers' Book.* There is a progression in complexity of language from Section to Section, and from Unit to Unit. Later Units incorporate the structures and vocabulary of earlier Units wherever possible, thus providing a constant element of repetition and revision.
There is a vocabulary list at the end of each Section, and a complete vocabulary at the back of the book. There are also Appendixes on *Irregular Verbs* and *SI Units of Measurement*. Tape-recordings of all the texts are available.
Book 2 contains twenty-two Sections, each providing enough material for 3–4 classroom periods. Teachers should find there is ample material for an academic year of 3–4 English periods per week.

Teaching a Section

Each Section begins with the oral introduction of new language items. Students then listen to the Listening Text. They may refer to the illustrations provided in their books, but the Listening Text does not appear there. After listening to the text, oral work can be continued with further language practice or comprehension questions. Next, students read the Reading Text in their books. The subsequent exercises, frequently illustrated, practise aural/oral skills as well as reading and writing. Since vocabulary is a highly important component of technical language, there are also a series of exercises to revise and enlarge students' vocabulary.

Contents of a Section

1. *Teaching Points* (Teachers' Book)
   These identify the language features (with examples) to be taught in each Section. This is followed by suggestions for their presentation and practice.
2. *Before Listening to the Text* (Teachers' Book)
   Full information on how to practise new language features is given, if necessary.
3. *The Listening Text* (Teachers' Book)
   New vocabulary and new language features are printed *in italics* the first time they appear in a Listening Text.
4. *After Listening to the Text* (Teachers' Book)
   The Listening Text is exploited for oral language practice.
5. *The Reading Text* (Students' Book)
   Each text has a picture or diagram(s) above it to aid comprehension and oral practice.
6. *Exercises* (Students' Book)
   Methods for using these, and a key wherever necessary, are given in the Teachers' Book.
7. *Language Notes* (Students' Book)
   Minor language features that arise in the Section are given in simple, often pictorial, form.
8. *Vocabulary List* (Students' Book)
   New words are grouped in columns of nouns, verbs, etc. Some Sections contain separate groups of technical terms.

How to Use Book 2

Since the course has a structural basis, we suggest that teachers deal with each Section thoroughly before passing on to the next.

Teaching Points
Read them through before starting a new Section. Look through the texts and the exercises to see how the points are presented and practised.

Before Listening to the Text
Oral drills are often used to present new language structures. Encourage students to use contracted forms when speaking. New vocabulary may be presented and reference made to the illustrations above the Reading Text.

The Listening Text
The texts are divided into short paragraphs to facilitate teaching. We believe teachers should have some licence to formulate their own methods for dealing with particular texts with particular students. Some teachers may even wish to allow their students to read quickly through the Reading Text at this stage in order to familiarize themselves with the general subject matter they are about to embark on. We do not recommend, however, that any significant amount of teaching time be spent on the Reading Text until after the Listening Text has been dealt with. Listening Texts are included as much for their information as for their language content. On both counts, they are *complementary* to the Reading Texts rather than repetitions.
Here is a generalized technique for dealing with Listening Texts which has been tried and tested:
1. Read out the first paragraph, pausing between sentences. (Students may use any relevant illustrations in their books or on the board to aid comprehension.) The meaning of individual words should be given where necessary. Use demonstrations or examples if at all possible. Translation can be used for 'difficult' words, but do not translate whole sentences.
2. Re-read the paragraph, sentence by sentence. Occasional repetition of phrases and sentences embodying new vocabulary and structures is recommended.
3. At the end of each paragraph treated in this way, read out the paragraph without pauses. Try to be a 'speaker' rather than a 'reader', i.e. refer to the printed text as little as possible.
4. After completing the final paragraph in this manner, read out the whole text again at normal speed. If you wish to ask short comprehension questions, or to expand on any information, explanations or language points in the text, pause between paragraphs.

After Listening to the Text
Any further preparatory work to be done orally before students turn to the Reading Text is included under this heading.

The Reading Text
Students should have few difficulties with comprehension if they have been able to follow the Listening Text. Encourage them to read whole paragraphs silently and quickly, with the help of a dictionary if possible. The following method for reading practice is recommended:
1. Students read each paragraph for general comprehension.
2. They read it again; raising any points that cause difficulty.

3. They read the paragraph once more for fluency practice. Reading at a natural speed can be encouraged by timing their reading, or asking them to read within a set time.
4. After completing the last paragraph, students should read through the whole text again before going straight on to the first exercise in the Section.

Exercises
Wherever possible, exercises refer closely to the topic of the Section. The main purpose(s) of many exercises is identified in the Teachers' Book by the rubrics *Reading* and *Vocabulary. Reading* designates exercises which either require students to refer back to the text or to read a further short text on a similar subject. *Vocabulary* is used for exercises which either revise or enlarge their vocabulary. New vocabulary introduced in an exercise is listed in the Teachers' Book. All exercises also bear the rubrics *Oral* and/or *Written*. This shows whether the exercise is to be used for oral or written practice, or both.
Some exercises contain substitution tables. Tables may also be written on the board during oral practice. Make sure your students know how to use them at an early stage in the course.
Many exercises are of the completion type. The following convention has been used in the Students' Book to mark spaces:
A line of 12 dots means a missing sentence.
A line of 8 dots means less than a sentence but more than one word is missing.
A line of 4 dots means a missing word.
Always consult the Teachers' Book before beginning an exercise so that the course can be used to its best advantage.

Language Notes
These generally cover minor language features not specifically covered in exercises. Refer your students to them at the appropriate time.

Vocabulary Lists
These are provided for study and revision. Encourage students to add notes or translations where necessary.

Appendixes
These are also provided for study and revision.

Vocabulary (at the back of the book)
Encourage students to use this like a dictionary. Meanings are not given but the reference after each word shows the students where it can be found in its original context.
NOTE: Words already listed in vocabularies in Book 1 are *not* listed again in Book 2.

Tape-Recordings

The Listening Texts in the Teachers' Book and the Reading Texts in the Students' Book are recorded on tape. In classrooms where the teacher is not a native speaker of English, the recordings of the Listening and Reading Texts will enable students to hear the voices of native English speakers. Recorded texts in the Teachers' Book are marked with the symbol ⊗. The recorded material can be used for:

1. Lesson preparation (checking the pronunciation of unfamiliar terms, etc.).
2. Classroom work (as an occasional substitution for the teacher's voice during listening practice, or for dictations).
3. Language laboratory work (comprehension).

Entry Test and Key

This test appears at the beginning of the Students' Book. It has been provided so that teachers can check their students' readiness to take up this course at Book 2 level.

The items in the test are all covered in Book 1, but they have been chosen for their general applicability to technical English. Thus the test can be used not only for students who have completed Book 1 – to test their retention of the material they have already encountered – but also as a diagnostic test for new students who may be starting a technical English course at the Book 2 level.

The test is marked out of sixty. Answers should be written on a separate sheet of paper.

It is difficult to give a precise pass-mark. However, we feel that any student *who has completed Book 1* and scores below forty marks needs some revision of Book 1 before attempting Book 2. For a student who has not previously used *A First Course in Technical English*, we would put the pass-mark somewhat lower (30–35 marks).

The test consists of:

A A Vocabulary Test

There are fifteen items, all taken from Book 1, Units 1–5.
One mark is given for each correctly written answer.

KEY

1 hacksaw 2 spanner 3 file 4 bucket 5 bottle 6 water tank 7 sack 8 thermometer 9 balance 10 telephone 11 copper 12 tin 13 mercury 14 iron 15 lead

15 marks

B A Listening Test

There are two parts. The first part consists of ten numerical items to be read out by the teacher and written down by the students:

KEY

1 *13* 2 *40* 3 *241* 4 *–38°C* 5 *97%* 6 *3·142* 7 *12½ mm* 8 *67⅔ cm³* 9 *19 g/cm³* 10 *$22 000 per tonne*

10 marks

The second part requires the students to draw a geometrical figure from the instructions you will read out. The material is taken from Book 1, Unit 2. A maximum of five marks should be given for the completed figure.

Here are the instructions:
Listen.
Follow these instructions and draw this figure.

1 Draw a small rectangle. Label the angles P, Q, R and S.
2 Mark the opposite and equal sides of the rectangle.
3 Draw a circle with centre Q and radius QR.
4 The circle cuts the side PQ at T. Mark the point T and join TR with a straight line.
5 Describe the shape TQR.
(It's a right-angled triangle with two equal sides.)

5 marks

C A Reading Test

The students must write down the correct word for each gap in the texts from the wordlist given. The materials are taken from Book 1, Units 5 and 8.

NOTE: There are more words in the lists than there are gaps in the texts.

KEY (First Test)

1 example	2 above	3 called	4 its
5 definite	6 same	7 m.p.	8 range

8 marks

(Second Test)

1 is	2 a	3 very	4 or
5 has	6 there	7 contains	

7 marks

D A Writing Test

This set of pictures and the wordlist are taken from Book 1, Unit 6 (p.88). A maximum of fifteen marks can be given for the eight written sentences.

KEY

1 Remove the filler plugs.
2 Fill the cell(s) with pure water.
3 Place the hydrometer in the (first) cell.
4 Press the bulb.
5 Release the bulb.
6 Read the scale.

7 Check the other (next) cell(s).
8 Replace the filler plugs.

15 marks

TOTAL – **60 marks**

UNIT ONE

Shapes and Conditions

SECTION A: Everyday Objects

Teaching Points

1 Revision

Unit One, Section A, marks the beginning of a new academic year. There are few new language features in it; apart from the new vocabulary in some of the exercises and the one main Teaching Point below, the emphasis is on revision from Book 1.

2 *-ed* words

The next three Units are chiefly concerned with the introduction of the Past Simple Tense and the Passive forms in English. This Section takes the opportunity to introduce words ending in *-ed*, *used as adjectives*. Students can thus concentrate on practising the pronunciation and spelling of the *-ed* suffix without encountering new grammatical features, which have yet to be learnt.

PRONUNCIATION

(a) [*ed*] = /ɪd/

EXAMPLES: calibrate – calibrated
corrode – corroded

(b) [*ed*] = /*d*/

EXAMPLES: curve – curved
use – used

(c) [*ed*] = /*t*/

EXAMPLES: crack – cracked
compress – compressed

SPELLING

All words simply add *-ed* except:

(a) Words ending in *-e*

EXAMPLE: corrode – corroded

(b) Words doubling the final consonant before the suffix

EXAMPLES: label – labelled
fit – fitted

(For further examples used in Unit 1, see LANGUAGE NOTE 2, at the end of Section B.)

(c) Words ending in *consonant+y*

EXAMPLES: vary – varied
supply – supplied

The Listening Text

There are no separate Listening Texts in Unit One. Use the Reading Text in the Students' Book for pronunciation practice.

NOTE: The three *-ed* words in the text exemplify each of the three different pronunciations of the suffix set out above.

After Listening to the Text

Make a list on the board of regular verbs which are familiar to the students. Add *-ed* to each and get the students to say them out loud. Let the students divide them into three groups according to how the *-ed* is pronounced.
Finally, pick some of the words and show with examples how they are used adjectivally.

Exercises

Exercise 1 (Oral/Written)

KEY

The fourth picture shows an electric fire.
The reflector is curved.

The fifth (next) picture shows a bottle of acid.
The bottle is labelled.

The sixth (last) picture shows a toolbox.
The toolbox is locked.

Exercise 2 (Oral/Vocabulary)

New Vocabulary: *finish(ed)* *paint(ed)* *cover(ed)* *insulate(d)* *compress(ed)*

Exercise 3 (Oral/Vocabulary)

New Vocabulary: *a shaft* *a slot* *a tooth* *a groove* *a scratch* *a dent* *a joint* *a thread*

Whereas previously *-ed* words have been formed from *verbs*, this exercise shows that they can also be formed from *nouns*.
Draw attention to the plural of *tooth* (*teeth*) (see LANGUAGE NOTE 1).

Exercise 4 (Oral/Written)

New Vocabulary: *badly* (*in*)*correctly* *half*

KEY

1 One tap is correctly fitted.
The other is incorrectly fitted.

2 One tank is fully covered.
The other is half covered.

3 One case is well labelled.
The other is poorly labelled.

4 One car is slightly dented.
The other is badly dented.

5 One pipe is too curved.
The other is not curved enough.

6 One instrument is exactly calibrated.
The other instrument is approximately calibrated.

Exercise 5 (Written)

Make sure students fill in the crossword in capital letters. All the clues (and answers) revise Book 1 material.

KEY

ACROSS		DOWN	
1 RAPID	21 TEEN	1 REFLEX	15 OFF
4 DRILL	22 EACH	2 ITS	18 DRY
7 TO	24 ALL	3 DO	19 CEMENT
8 IE	25 TYRE	4 DI	20 STRESS
9 FINS	26 CA	5 RED	23 HCL
10 HOT	28 OR	6 LARGER	25 TRI
12 DOOR	29 TOOLS	11 OCTAGONAL	27 AS
16 THAT	30 FILES	13 CHARGED	28 OF
17 FIREGUARD		14 BALANCE	

SECTION B:
A Light Bulb and Socket

Teaching Points

1 *-ed* **words** (cont.)
Section B takes the use of *-ed* words in English a stage further.
A sentence such as:

The conductor is insulated

which might have occurred in Section A is here extended to:

The conductor is insulated inside a glass mount.

The Reading Text exemplifies this type of sentence, which is frequently used to describe assemblies of machine parts, etc., and the usage is practised in all five exercises in this Section.
Exercises 7 and 8 further exemplify and practise *-ed* words formed initially from nouns.
Exercise 10 attempts to distinguish between the above type of description of a finished assembly:

EXAMPLE: *The pulley* is fitted *onto the shaft*
and a description of *how* assemblies *can* be fitted together as exemplified by the exploded diagrams with which technical manuals abound:

EXAMPLE: *The pulley* fits *onto the shaft.*

Before Listening to the Text

Explain and practise orally the new vocabulary in the two diagrams above the Reading Text. Encourage questions (in English) about the words and the diagrams.

The Listening Text

Again, there is no separate Listening Text. Allow your students to examine the two diagrams before slowly reading the text to them.

After Listening to the Text

After treating each paragraph as above, elicit oral sentences from your students by using *-ed* words as key words. Pay particular attention to the use of the correct preposition in each case.

TEACHER: hooked	STUDENT: The conductor(s) is/are hooked over the filament.
filled (etc.)	The bulb is filled with argon gas. (etc.)

Exercises

Exercise 6 (Oral)

New Vocabulary: *mixed* *a washer*

KEY

1 The bulb is filled with argon gas.
2 The handle is covered with rubber.
3 The wire is insulated with plastic.
4 The car is fitted with a radio.
5 The joint is sealed with a washer.
6 The pipes are painted with lead oxide paint.
7 The sand is mixed with coarse aggregate.

Exercise 7 (Written)

New Vocabulary: *terminals*

KEY

1 The mirror is screwed onto the car.
2 The wires are soldered (on) to the terminals.
3 The bell is nailed (on) to the workbench.
4 The kettle is plugged into the socket.
5 The socket is fixed (on)to the wall.
6 The headlights are connected to the battery.
7 The valve is bolted (on)to the pipe.
8 The pylon is concreted into the ground.

Exercise 8

New Vocabulary: *a castle nut* *a key* *a clevis pin* *a split pin* *a taper pin* *a dowel pin* *a circlip* *a hose clip* *a rivet* *a pop rivet* *a clamp* *a stud* *a lock washer* *a clip* *a set-screw*

KEY

1 The patch is riveted (on) to the tank.
2 The castle nut is pinned onto the bolt.
3 The pulley is pinned (on)to the shaft.
4 The water hose is clipped over/onto the pipe.
5 The pulley is keyed (on)to the shaft.
6 The screwdriver is clipped onto the wall.

Exercise 9 (Written)

Give students plenty of time to make up correct sentences and write them down. Then ask individuals to read out what they have written. Other students can judge whether each sentence is permissible or not. This technique will also give each person the opportunity to add to his list from the sentences read out. Each of the nouns in the second column of the table can be used in several different sentences.

EXAMPLE: *filament*

The filament is	made of a tungsten wire. sealed inside the glass bulb.

Exercise 10 (Oral/Written)

New Vocabulary: *a torch* *a radio* *a pen* *a stand*

The distinction between (a) and (b) type sentences is described in the Teaching Points.

NOTE: The difference in usage between *fitted* and *fixed* is difficult to distinguish and therefore should be avoided.

UNIT TWO | **Steam Engines**

SECTION A: Newcomen's Engine

Teaching Points

1 **Past Simple Tense** (Regular Verbs)
The use of the Past Simple in Section A is confined to the affirmative form of regular verbs.
This tense is easier to present to the learner than the Present Simple.
Explain to the students that:

(i) The tense is used (with rare exceptions not dealt with here) for describing events, objects, etc. which quite clearly belong to the past.
(ii) The verb form, either spoken or written, remains the same in all persons, singular and plural.
(iii) The *-ed* suffix is used on all regular verbs.

2 *Was/Were* and *Had*
These are the only irregular verb forms in this Section. They are all practised in Exercise 2. Point out that *was* and *were* have contracted forms in the negative in speech (*wasn't* and *weren't*).

3 **Possessive** *'s*
Possessive *'s* is most frequently used with personal names, as will be seen several times in the next two Units. It is also frequently used with other names or professions.

EXAMPLES: The engineer's materials.
The Earth's surface.

It is seldom used with other nouns.

EXAMPLE: The movement of the piston.
Not usually – The piston's movement.

Before Listening to the Text

1 Some dates appear in this Section. Explain how they are realized in speech.

EXAMPLES: 1800 – eighteen hundred.
1804 – eighteen hundred and four.
1850 – eighteen-fifty.
1977 – nineteen seventy-seven.

2 Make sure that the irregular forms *was/were* and *had* are known. Use some contrasting examples with dates, such as:

In 1980, steam engines are quite complex machines.
In 1850, steam engines were very simple machines.

In 1980, the fuel is often oil.
In 1850, the fuel was always coal.

In 1980, steam engines have a lot of power.
In 1850, steam engines had little power.

3 Look at the diagram of Newcomen's engine with your students. Identify and practise orally the parts of the machine which correspond to the labels A–N.

4 Introduce possessive *'s*. Remind the students that names have capital letters.

The Listening Text

Heat converts water into steam. Steam has a much greater volume than water. At normal air pressure, 1 cubic centimetre of water expands into 1600 cubic centimetres of steam. This provides an extremely important *source* of power.

In 1850, there *were* no petrol engines and no diesel engines. Steam *was* the only important source of mechanical power. The first steam engines *weren't* at all *complex*. They were usually large and heavy and they *had* comparatively little power. However, their *design* was quite *similar* to later engines.

After Listening to the Text

Ask students to skim through the Reading Text and locate all the *-ed* words. Write them on the board as they are called out (there are seventeen of them) and practise their pronunciation. Get students to write them out in three groups according to their pronunciation – (/ɪd/), (/d/) or (/t/).
Any explanations about the Past Tense and its use in English should be made at this point.

Exercises

Exercise 1 (Oral/Reading)

KEY

1 True.
2 False. The water boiled in a copper boiler.
3 True.
4 True.
5 True.
6 False. The air pressure forced the piston to the bottom of the cylinder.
7 True (see diagram).
8 False. It worked at a very low speed.

Exercise 2 (Oral/Written)

New Vocabulary:	*a minute*	*a year*	*nowadays*
	powerful	*later*	*complex*

In Nos. 6–10, practise with your students the contracted oral forms *wasn't* and *weren't*.

KEY

1 Newcomen's engine had a copper boiler.
2 Newcomen's engine had wooden parts.
3 Newcomen's engine had a coal furnace.
4 Newcomen's engine had (only) one cylinder.
5 Newcomen's engine had a beam.
6 Newcomen's engine was large. Later engines were not as large.
7 Newcomen's engine was slow. Later engines were not as slow.
8 Newcomen's engine was expensive to operate. Later engines were not as expensive to operate.
9 Newcomen's engine was not very complex. Later engines were more complex.
10 Newcomen's engine was not very strong. Later engines were stronger.

Exercise 3 (Oral/Written)

Ask your students to write out the description first as a single paragraph. They will probably have to refer back to the Reading Text to complete it. Asking them to read out what they have written forms the oral part of the exercise.

. . . Steam entered the cylinder through a valve. Steam pressure forced the piston up the cylinder. The beam turned on its centre point. The movement of the beam lowered a bucket into a well. Water (then) entered the cylinder through a valve. Air pressure forced the piston down the cylinder. The movement of the piston turned the beam on its centre point. The movement of the beam (also) raised the bucket from the well.

Exercise 4 (Oral/Written)

Again it would be useful practice to write out the ten sentences as a single paragraph.

SECTION B: Watt's Engines

Teaching Points

1 **Past Tense** (Question form)

Inform students that the question form is in every way analogous to that of the Present Simple, with *does/do* becoming *did*.

EXAMPLES: *Did* Watt's engine have a boiler?
How *did* Watt improve engine design?

2 *And so*

We have used *and so* in the construction:

Copper is very malleable *and so* is lead (Book 1, Unit 7).

In the present Section it is used in a different construction.

EXAMPLE: The furnace required less fuel *and so* the engine was more efficient.

This usage is contrasted with *because* in Exercise 7.

3 *How* (+**verb**) . . . . ?

Up to now, students have only encountered *How+adjective* . . . . ? They will need to be shown that there is a semantic as well as a grammatical difference between the two uses of *How* . . . . ? Practice with *How+verb* is given in Exercise 5.

4 *He* and *His*

These two words first appear in this text. Introduce them in a simple context.

EXAMPLES: Watt constructed *his* first engine in 1765.
His engine was more powerful than earlier engines.
He fitted a steam jacket around the cylinder.

Then contrast *he/his* with *it/its*.

EXAMPLES: A hammer is a tool. *It* has a steel head. *Its* head is made of steel.
Watt was an engineer. *He* constructed engines.
His engines were used for pumping water.

You can take this opportunity to introduce *she/her* and *they/their* (for persons) in a similar way.

Before Listening to the Text

Introduce and practise the engine parts which correspond to the labels A–H in **fig. (b)**.
Then introduce the Past Simple question form by asking the questions below. Students should be looking at **fig. (a)** and give short answers.

TEACHER:

Did Newcomen's engine have Watt's	a boiler?	(Yes, it did./ No, it didn't.)
	a condenser?	
	a closed cylinder?	
	a water jet?	
	an air pump?	
	a beam?	
	a steam jacket?	

The Listening Text

(*To introduce* efficient/efficiency.)

Petrol, diesel, coal, gas, etc. are fuels. Fuels are used for producing heat *energy*. This heat energy provides the engine power. However, not all the heat energy becomes engine power. Some engines use a lot of fuel and provide very little power. That's to say, these engines aren't very *efficient*. The *efficiency* of an engine is determined by the ratio of fuel to power. Efficiency is expressed as a percentage. Newcomen's engine had only 1% efficiency. This means that only about 1% of the heat energy from the fuel produced power.

Watt's engines were more powerful and they used less fuel. In other words, they were more efficient.

(*To introduce* and so, he *and* his.)

How did Watt's engines become more efficient? Look at **figure a**.

In Newcomen's engine, heat escaped because the top of the cylinder was open.
In Newcomen's engine, the top of the cylinder was open *and so* heat escaped.

Newcomen's engine was less efficient because the top of the cylinder was open.
The top of the cylinder was open and so the engine was less efficient.

Watt's engines were more efficient because the cylinder had a steam jacket.
He fitted a steam jacket around the cylinder and so *his* engines were more efficient.

Watt's engines were also more efficient because they had a *condenser*.
He fitted a *separate* condenser and so his engines were more efficient.

After Listening to the Text

1 Now would be a good time to practise personal pronouns and adjectives (see Teaching Point 4 at the beginning of this Section).

2 Introduce *How did . . . . ?* questions. *Either* use a simplified version of some of the questions in Exercise 5 *before* students read the text, *or* use Exercise 5 as an oral exercise *after* students have read the text.

Exercises

Exercise 5 (Oral/Written/Reading)

Answering these questions necessitates a good grasp of the text. Allow time for oral discussion of relevant points in the text and of how answers should be phrased. The questions are more 'open-ended' than the previous ones and so the KEY is in note form only.

KEY

1 By escaping through the cylinder wall and the top of the cylinder. (Students may also notice that energy escaped through the use of a water jet inside the cylinder.)
2 By the use of a steam jacket (and the top of the cylinder was closed); and the use of a separate condenser.
3 By using two valves.
4 The use of steam pressure for both the up and down stroke.
5 Use of a separate rod (F).
6 Via the same rod.

Exercise 6 (Written)

If oral practice is desired, it can be attempted after the writing has been completed.

KEY

1 The steam entered the cylinder and forced the piston down the cylinder.
2 The piston travelled down the cylinder and lowered (one end of) the beam.
3 A rod closed one valve and opened the other (valve).
4 Steam entered (the bottom of) the cylinder and forced the piston up the cylinder.
5 The piston travelled up the cylinder and raised (one end of) the beam.
6 The rod closed one valve and opened the other (valve).
7 A (third) valve opened and removed the steam from the cylinder (to the condenser).
8 (Another) piston travelled up the air pump and emptied (the water from) the condenser.

Exercise 7 (Oral/Written)

New Vocabulary: *vary*

KEY

1 The cylinder temperature varied *and so* Newcomen's engine had poor efficiency.
2 The steam condensed inside the cylinder *and so* the cylinder temperature varied.
3 The steam pressure was not high *because* steam escaped through the top of the cylinder.
4 A large amount of steam was used for raising the cylinder temperature *because* the cylinder temperature varied considerably.
5 Watt's engines were more powerful *because* steam pressure supplied the power for both the up and the down strokes.
6 The steam condensed rapidly in the condenser *and so* Watt's engines had a higher speed.
7 The steam condensed outside the cylinder *and so* the cylinder temperature remained stable.
8 The cylinder remained hot *because* there was a steam jacket around the cylinder.
9 Watt's engine was four times as efficient as Newcomen's engine *because* (or *and so*) the furnace used only 25% as much fuel.

Exercise 8 (Oral)

New Vocabulary: *a loss* *consumption*
a day *rapid*

KEY

1 The fuel consumption / The consumption of fuel was smaller.
2 The engine design / The design of the engine was more complex.
3 The steam pressure / The pressure of the steam was higher.
4 The piston stroke / The stroke of the piston was more powerful.
5 The cylinder temperature / The temperature of the cylinder was more stable.
6 The beam movement / The movement of the beam was more rapid.
7 The engine power / The power of the engine was greater.
8 The engine efficiency / The efficiency of the engine was greater.

SECTION C: Parsons' Turbine

Teaching Points

1 Irregular Verbs (Past Simple Tense)

Many of the most common verbs in English are irregular. This last text presents nine irregular lexical verbs in their Past Tense form.
A list of irregular verbs is to be found in Appendix One.

Inform students that:

(i) Irregular verbs (apart from *be*, *have* and *do*) only show their irregularity in the Past Tense (and the Passive).
(ii) In negative and question forms, the verb reverts to its original form.

EXAMPLES: Turbines *became* very efficient.
Piston engines did not *become* very efficient.

Rotors *drove* the shaft.
What speed did they *drive* it at?

Past Tense questions with irregular verbs are practised in Exercise 9.
Past Tense negatives are practised in Exercise 10.

2 *What did . . . . do?*

The verb *do* has only previously appeared as an auxiliary in questions and negatives. It can, of course, also be a lexical verb.

EXAMPLE: The student read the text and then *did* the first exercise.

It is commonly used in the above question form (or in *What do/does . . . . do?*). It can be a confusing construction for learners because of the double occurrence of *do*, and so it has been singled out for practice in Exercise 12.

Before Listening to the Text

1 Explain the relevant features of irregular verbs (see above). Tell students to open their books at Appendix One. Name the nine irregular verbs used in the text for them. Give their meanings by translation. Let students identify the verbs (and their groups) in the Appendix.

Practice by oral repetition:

become – became
build – built
come – came
cut – cut
drive – drove
give – gave
go – went
put – put
run – ran

Then, with books closed, see if the students can remember the Past Tense forms (and the meanings) as you read out the Present Tense forms.

2 While students refer to the text illustrations of the various items, go through orally the engine parts with them.

The Listening Text

Nowadays, steam *still* provides a very important source of power. However, *modern* engines are usually a different type of steam engine. These are called *turbines*. Turbines don't have pistons. They are much more efficient than piston engines and they work at much higher speeds.

Nowadays, steam engines are generally used for *generating* electricity. A machine for generating electricity is called a *generator*. The power of the machine is normally measured in watts (W) or kilowatts (kW). The generator produces electrical energy. This is measured in joules (J) or kilowatt hours (kW h). Steam pressures are measured in kilopascals (kPa). (*Refer to Appendix Two if further explanation of these units of measurement becomes necessary.*)

In 1884, Parsons designed one of the first electrical generators. Before then, it was impossible to construct efficient generators because they needed an engine speed of more than 1000 *revolutions per minute* (*rpm*). Piston engines *didn't* have the ability to *run* at this speed. They all *ran* at less than 150 rpm. Efficient generators only *became* possible with Parsons's new turbines. His first turbine *drove* the generator at a speed of 18 000 rpm.

After Listening to the Text

Explain and practise the Past Tense negative form. Use the following table:

Newcomen's engine Watt's engine	had did not have	a wooden beam a steam jacket a condenser an efficiency of 4% an air pump a water jet

Exercises

Exercise 9 (Oral/Reading)

KEY

1 Fourteen (seven in each cylinder).
2 Metal rings (with slots in them).
3 Yes.
4 No.
5 18 000 rpm.
6 It pushed against the slot walls.
7 No.
8 Through the shaft.
9 100 V.
10 (a) Engine speed too high.
(b) Exhaust steam escaped under pressure (i.e. not all the steam energy became engine power).
11 (a) Steam condensers.
(b) Larger rotors, larger dynamos, gearboxes.
12 More than 90%.

Exercise 10 (Reading/Written)

Regular verbs		*Irregular verbs*	
supply	force	put	drive
enter	travel	cut	go
expand	push	become	have
consist	turn	give	run
slope	use	come	

(Oral)

KEY

1 Ran (did not run).
2 Put (did not put).
3 Cut (did not cut).
4 Gave (did not give).
5 Came *or* went (did not come *or* go).
6 Drove (did not drive).
7 Went (did not go).
8 Became (did not become).

Exercise 11 (Oral/Written)

KEY

1 The boiler supplied the steam at a pressure of 550 kPa.
2 The steam entered the engine casing at high speed.
3 The steam expanded rapidly in opposite directions along the two cylinders.
4 The steam forced its way through the slots.
5 The steam pushed all the rotors in the same direction.
6 The rotors turned the shaft at 18 000 rpm.
7 The shaft drove the dynamo at the same speed.
8 The dynamo produced a voltage of 100 V.

Exercise 12 (Oral)

Use the example at the top of the exercise for initial practice of this unfamiliar question form.

TEACHER: What did the boiler / rotors / shaft / dynamo do?

STUDENTS: It supplied the steam.
They turned the shaft.
(etc.)

Now ask students to identify the MAIN PARTS (a–e) of Newcomen's engine and write them in the box provided. They are: *the boiler*, *the piston*, *the beam*, *the pump rod*, and *the water jet*.

Students can now take part in asking questions as well as answering them.

What did	the boiler the piston the beam the pump rod the water jet	do?	It supplied the steam. It turned the beam. It raised and lowered the pump rod. It raised and lowered the bucket. It condensed the steam.

Continue in the same way with Watt's engine.

(a) The boiler – it supplied the steam.
(b) The steam jacket – it heated the cylinder.
(c) The piston – it raised and lowered the beam.
(d) The beam – it raised and lowered the pump rod.
(e) The condenser – it condensed the steam.
(f) The rod – it opened and closed the valves.
(g) The air pump – it emptied (the water from) the condenser.

UNIT THREE

Laboratory Experiments

SECTION A: Calculating Density [i]

Teaching Points

In this Unit we introduce two very specific uses of the Present Simple Tense, both very commonly used wherever students of science and technology are being taught.

1 Present Simple Tense Active (Demonstrations)

First we have the use of the Active Tense in demonstrations. Whenever a teacher or demonstrator gets up before his students to show them an actual experiment or procedure, he is going to say:

EXAMPLE: First, I *measure* the dimensions of . . . .
and *not* First, I *am measuring* the dimensions of . . . .

He is equally likely to say:

EXAMPLE: First, *I* measure . . . .
First, *we* measure . . . .
First, *you* measure . . . .

In the context of demonstrations, the choice of personal pronouns used (*I*, *we* or *you*) is generally a matter of an individual's habits and does not alter the meaning of the sentence at all.

NOTE: This is the first use of the Present Simple Tense in this course other than in the 3rd person (with *he*, *she*, *it* or *they*). Point out that the 1st and 2nd person forms do not have the *-s* suffix of the 3rd person singular form.

2 Present Simple Tense Passive (Descriptions)

The Passive Tense will almost always be used for a *written* description of the same experiment or procedure.
Whereas the demonstrator will say:

EXAMPLE: We weigh the block . . . .

a written description will read:

EXAMPLE: The block is weighed . . . .

Throughout this Unit, the Simple Passive forms are purposely contrasted with the Active forms more likely in speech. It is hoped that this technique will enable students to grasp the semantic differences between Active and Passive from the outset. As for formal differences, it is hoped that these will cause few problems. Although students have not encountered the Passive proper before this Unit, they have encountered its form in:

is/are + *used/made* (Book 1)

and

is/are + *-ed* words (Book 2, Unit 1)

The Listening Text

Use the series of pictures and the substance of the Reading Text to simulate an actual demonstration. An example is provided on tape. The first paragraph is an introduction such as any teacher might use before this experiment. Notice the word *mass*. Mass is the correct scientific term for what a non-scientist would call 'weight' in this context.

Written	*Spoken*
$\frac{m}{v}$	m over v
g/cm^3	grams per centimetre cubed
kg/m^3	kilograms per metre cubed
ρ	rho

NOTE: g/cm^3 can also be written gcm^{-3} ($kg/m^3 = kgm^{-3}$).
For further examples, see Appendix Two.

The actual experiment should be simulated using *Active forms only*. Point out at a suitable moment that *I*, *we* or *you* are equivalents. Once the demonstration is thoroughly understood and any oral questions are over, repeat the experimental procedure aloud, giving a Passive as well as an Active sentence for each of the six pictures. An explanation of the different uses of Active and Passive in this context will lead directly on to Exercise 1.

Exercises

Exercise 1 (Oral/Written)

KEY (Written Forms)

1 A ruler is needed .... for this experiment.
2 The block is measured .... with the ruler.
3 The volume is calculated .... with the formula lwh.

4 The block is weighed . . . . on the balance.
5 The density is calculated . . . . with the equation $\rho = \frac{m}{v}$.
6 The equation $\rho = \frac{m}{v}$ is used . . . . for calculating density.

Exercise 2 (Written)

Point out the usual difference between instructions (Imperatives) and demonstrations (Simple Tense Actives). The picture of a fire extinguisher refers back to Book 1, Unit 4, where instructional language was originally introduced.

Exercise 3 (Reading/Writing)

Allow students ample time to read this new but related description of an experiment.

NOTE: *First* and *finally* are new items of vocabulary.

In the written part of the exercise, students have only to copy out what they have just read and substitute the correct numerical values for the block of lead.

$$m = 113{\cdot}4\text{ g} \quad v = 2\times2\times2{\cdot}5 = 10\text{ cm}^3 \quad \rho = \frac{113{\cdot}4}{10} = 11{\cdot}34\text{ g/cm}^3$$

Exercise 4 (Written)

KEY *Experiment 2 A*

To calculate the density of a liquid.

A balance is needed.
A density bottle (100 cm^3) is also required for this experiment.

1 First the bottle is washed with alcohol.
2 The empty bottle is weighed on the balance.
(Its mass is 10 g.)
3 Then the bottle is filled with the liquid.
4 Next the full bottle is weighed on the balance.
(Its mass is now 100 g.)
5 The mass of the liquid is calculated with the equation
$m_3 = m_2 - m_1 = 100 - 10 = 90$ g.
6 Finally its density is calculated with the equation
$\rho = \frac{m}{v} = \frac{90}{100} = 0{\cdot}9\text{ g/cm}^3$.

The mass of the benzene is $0{\cdot}9$ g/cm^3
The mass of the water is $1{\cdot}0$ g/cm^3
Benzene is lighter than water. That is why petrol and even oil float on water.

Exercise 5 (Oral)

This exercise affords a good opportunity to introduce Appendix Two.

SECTION B: Calculating Density [ii]

Teaching Points

1 Simple Passive (cont.)

Section B provides further practice of the Passive in the same context. Several new features are introduced:

A Passive Questions (Exercise 6)

These should cause no learning problem. Passive questions are formed according to rules already learnt. Draw attention to the difference between the spoken and written versions of the same question form as exemplified in LANGUAGE NOTE 7.

B Passive with Adverbs

The adverb is usually placed directly after *is/are*.

EXAMPLES: Its mass is *accurately* weighed.
The vessel is *then* filled with water.

C Past Simple Tense

In the Passive, *is/are* is replaced by *was/were*.

2 Reports

Reports on laboratory experiments are generally written up in a particular and very restricted style. They are naturally written in the Past Tense, so students are introduced to the Simple Past Passive while they practise this important technical language activity.

The normal format for reports has four parts:

Aim: Stated as concisely as possible.

Equipment: A simple list of all the items of equipment needed for that experiment.

Procedure (sometimes called *Method*): An exact description of all the steps taken during the experiment.

Conclusion: A concise statement of any results from the experiment which relate to the aim.

Exercises 7 and 8 introduce and practise report-writing. Exercise 9, which contains two further experiments, can either be written up as descriptions (like the Reading Text), or as reports.

3 **Quantities** (Materials)

Students have already been taught that mass nouns cannot be pluralized with the *-s* suffix (Book 1, Unit 1). In order to quantify *wood*, *sand*, *glass*, etc., one method is to use the construction:

a/two/some . . . . of

Students are already familiar with *a lot of* (Book 1, Unit 3) and *a . . . . amount of* (Book 1, Unit 8), but there are many more such phrases which describe approximately the shape, size or amount of different materials. Exercise 10 introduces and practises some of the more common ones.

4 *Let/Allow*

Although these words are generally synonomous in most contexts, they have separate usages. Whereas *let* is used informally and in speech, *allow* is more usual in formal or written English. Moreover, *let* is followed by an infinitive without *to*, whereas *allow* takes *to* plus infinitive.

EXAMPLES: *Let* the solid *sink* . . . . (spoken/informal)
The solid is *allowed to sink* . . . . (written/formal)

The Listening Text

Simulate the experiment described in the Reading Text, using spoken language (instructional or demonstration language) in the same way as suggested for Section A. An example is provided on tape.
It may be more convenient to sketch on the board rough pictures of the various stages of the experiment while you are 'demonstrating', rather than to refer students continually to the illustrations in their books.

After Listening to the Text

Use the questions of Exercise 6 orally to check comprehension of the experiment. Students may give short answers.

NOTE: Question 3: Why *isn't* a ruler required? (Because it's impossible to measure an irregular piece of metal with it.)

Exercises

Exercise 6 (Written/Reading)

Students have already answered these questions orally before they read the text. Now they should write out the answers in formal sentences.

Exercise 7 (Written/Reading)

New Vocabulary: *a report*

Allow your students to refer back to the Reading Text if they have any major difficulties with these sentences.

Exercise 8 (Written/Reading)

New Vocabulary: *a conclusion* *probably*

Most of the report can be copied out exactly. Make sure that students become aware of the format for report writing (as set out in the Teaching Points) and of its layout.

Experiment 3 A $\rho = \frac{722}{95} = 7{\cdot}6\ \text{g/cm}^3$

Thus, the conclusion should state that the piece of metal is probably cast iron.

Exercise 9 (Oral/Written)

It is better to ask students to describe the experiments orally first, to ensure that they see how the result is arrived at.
In the first experiment, the difference in the two levels of water is noted instead of, as previously, letting the displaced water overflow into a measuring cylinder.

$$\rho = \frac{360}{72} = 5\ \text{g/cm}^3$$

In the second experiment, the wood must be attached to a piece of metal or it will not sink and displace the correct amount of water. The water displaced by the *metal*, however, must be discounted in the calculations, and so it is removed by putting the metal in the water first and letting the displaced water spill out.

$$\rho = \frac{54}{81} = 0{\cdot}67\ \text{g/cm}^3$$

NOTE: Wood is obviously less dense than water, so its density is less than 1 g/cm³.

Exercise 10 (Oral/Vocabulary)

KEY (from left to right)

a length	a reel (of cotton)	a piece	a sheet (of glass)
a piece	a reel (of wire)	a piece	a strip (of leather)
a bucket	a grain	a pile (of sand)	
a piece	a plank	a stack (of wood)	

UNIT FOUR | **Industrial Chemistry**

SECTION A: The Production of Copper

Teaching Points

1 **Simple Passive** (Processes)

In the last Unit, stylistic and grammatical differences between Active and Passive equivalent sentences were contrasted.
When the processes common to chemical and manufacturing industries, etc. are described, the Passive is also frequently used.
However, there are often no Active equivalents to sentences like:

The ore is crushed into fine grains.
The minerals are separated from the rock particles.

Moreover, there are no differences between spoken and written language based on Active/Passive equivalents in this new context.

2 ***-er/-or*** **Suffix**

Nouns are commonly formed from verbs by adding the above suffix to denote the agent that performs an action. It may be a person (*teach-er*) or a mechanical object (*crush-er*). In this Section, the students' attention is drawn to the use of this suffix to denote machines of one type or another (see Exercise 4).
Students should try to memorize which words end in *-er* and which in *-or* (see LANGUAGE NOTE 8).

3 **Chemical Terminology**

The terminology used in this Section can be briefly explained thus:

An *ore* is what is mined from the ground. It probably contains several minerals as well as a lot of waste material which must be discarded.

A *concentrate* is what is left of the ore after most of the waste material has been removed. It is therefore *rich* in minerals.

A *mineral* is a chemical compound containing a metal. The pure metal is extracted from its mineral forms. In this Section, the minerals mentioned are:

copper sulphide (Cu_2S)
copper oxide (CuO)
iron sulphide (FeS)
iron silicate ($FeSiO_3$)

NOTE: (i) That the metal is always mentioned first in the name.
(ii) Minerals ending in *-ide* only contain a metal plus *one* other element.
(iii) Minerals ending in *-ate* contain a metal plus *more than one* other element (including oxygen).

Before Listening to the Text

Ask students to look at the illustrations. Explain to them that the text describes how copper is produced from its ores. Each picture illustrates one step in the process. Discuss the various machines illustrated and practise their names. Students need a general idea of what each machine does – the oxidizer adds oxygen to the concentrate; the electrolytic bath is the same as an electrolytic cell (see Book 1, Unit 6).

The Listening Text

Metals are usually *extracted* from *ores*. The ores are extracted from the ground. An ore's a mixture of *minerals*, rocks and other materials. Minerals are chemical compounds. They always contain one metal element plus other non-metal elements.
(*Give some examples of minerals, using the blackboard. You can also draw a simple chart to illustrate the next paragraph.*)

Pure copper's extracted from ores. The production process is quite complex. First the ore's made into a *concentrate*. Most of the non-minerals are removed. The concentrate's *rich* in minerals. The copper minerals are then extracted from the concentrate. Finally the minerals are *transformed* into pure copper.

There are two main types of copper ore – copper sulphide ores and copper oxide ones. Copper sulphide ores are much more common. The production process for copper sulphides is described in the Reading Text. Exercise 3 describes a different process used with copper oxides.

After Listening to the Text

Write some chemical compounds on the blackboard. Practise their names and formulae.

EXAMPLES: Cu_2S – copper sulphide – C–U–2–S
$FeSiO_3$ – iron silicate – F–E–S–I–O–3

(Many other examples can be found – for example, in Book 1, Appendix Two.)
Carry on by asking:

What elements does	copper sulphide sulphur dioxide sodium chloride (etc.)	contain?	copper and sulphur. sulphur and oxygen. (etc.)

After students have *read* the text, you can also use the following question pattern before going on to Exercise 1.

TEACHER:	What does the	crusher oxidizer converter flotation cell furnace electrolytic bath	do?	It crushes. . . . It adds oxygen. . . . (etc.)

Exercises

Exercise 1 (Oral/Reading)

What happens . . . .? was first used in Book 1, Unit 6. It is a very common question form which may require further practice. Students may ask as well as answer the questions.

Exercise 2 (Written)

KEY

1 The ore is crushed into fine grains.
2 The copper-rich grains are separated from the rock particles.
3 Hot air is passed through the oxidizer.
4 SO_2 gas is allowed to escape (from the oxidizer).
5 Silica is added to the concentrate (inside the furnace).
6 Iron sulphide (FeS) is transformed into iron silicate ($FeSiO_3$).
7 Iron silicate is removed (from the surface of the concentrate).
8 The copper sulphate plus oxygen is converted into pure copper plus sulphur dioxide gas.
9 The liquid copper is allowed to cool (into blocks).
10 The copper anodes are placed in an electrolytic bath.
11 An electric current is passed through the electrolytic bath.
12 The copper molecules are (eventually) deposited on the cathode.

Exercise 3 (Written)
It may be wise to go through this alternative process orally first.

KEY

1 The copper ore is crushed into fine grains.
2 The copper ore is placed into concrete tanks.
3 Dilute sulphuric acid ($H_2SO_4 + H_2O$) is added to the ore.
4 The sulphuric acid plus copper oxide is converted into copper sulphate plus water.
5 The dilute copper sulphate is removed from the tanks.
6 The copper sulphate is transferred to an electrolytic bath.
7 An electric current is passed through the bath.
8 The copper molecules are (eventually) deposited on the cathode.

Exercise 4 (Oral/Vocabulary)
New Vocabulary: *a projector* *a mixer* *a scriber* *a refrigerator* *a typewriter*

KEY

1 A ruler (ruling lines).
2 A fire extinguisher (extinguishing fires).
3 A boiler (boiling water).
4 A reflector (reflecting heat).
5 A radiator (radiating heat).
6 A generator (generating electricity).
7 A wiper (wiping windscreens).
8 A projector (projecting films).
9 A mixer (mixing concrete).
10 A scriber (scribing, or writing on metal).
11 A refrigerator (refrigerating food).
12 A typewriter (typing letters).

SECTION B: The Production of Aluminium

Teaching Points

1 Simple Passive for Processes (cont.)

The use of the Passive for describing processes is continued in this Section in more complex sentences. Some irregular verbs are included in the text and Active/Passive equivalent sentences are again contrasted in Exercise 6.

Verbal adjectives (ending in *-ed*) were introduced in Unit 1. In this Section, they are used attributively (i.e. before the noun) for the first time. Exercise 8 is devoted to practising this point.

2 Two Suffixes

A *-ize*

Verbs denoting processes can sometimes be formed from nouns by adding *-ize*. *Oxidize* was used in Section A; there are two further examples in the present Section (see LANGUAGE NOTE 10).

B *-ite*

The names of many ores end in *-ite*. Section B has *bauxite* and *cryolite*. Section C contains *haematite* and *magnetite* which are iron ores.

The Listening Text

Aluminium's the most common metal in the *Earth's* surface. About 8% of the Earth's *crust* is composed of aluminium. However, aluminium's never *found* in its pure state. Like copper, it's extracted from ores. Because aluminium ores are very common, aluminium is a comparatively cheap metal.

The most common type of aluminium ore is called *bauxite*. It was *named* after a *town* in *France*. Bauxite usually contains a mixture of aluminium oxide, iron oxide, silica and water. Silica's the common name for silicon oxide (SiO_2). In the same way, aluminium oxide has a common name. It's called *alumina*. Both silica and alumina are found in cement.

The production process for alumina is *given* in the Reading Text. Pure aluminium is then refined from alumina in an electric furnace. This process is *shown* in Exercise 7.

After Listening to the Text

Point out the irregular Passive forms *found*, *given* and *shown* as they occur. Use Appendix One to demonstrate that irregular verbs often have two different irregular forms (one for the Past Simple and one for the Passive). Ask your students to skim through the Reading Text to find the irregular forms there (see Exercise 5). They can also identify from the Appendix the Passive forms for other verbs they have only come across up to now in the Present or Past Active forms.

Exercises

Exercise 5 (Written/Reading)

KEY

1 True.
2 True.
3 True.
4 False. The water content is extracted. . . .
5 True.
6 True.
7 False. The dissolved sodium aluminate passes through a filter.
8 False. The undissolved impurities are left inside the filter.
9 True.
10 False. Alumina (aluminium oxide), *not* aluminium.

Exercise 6 (Oral/Written)

KEY

1 The bauxite is crushed between rollers.
2 The bauxite is powdered with heavy steel balls.
3 The aluminium oxide is dissolved in sodium hydroxide.
4 The undissolved impurities are removed in the filter.
5 Larger aluminium oxide crystals are formed from the aluminium hydroxide crystals.
6 The crystals are converted into pure alumina in the drier.

Exercise 7 (Reading/Vocabulary)

The Reading Text dealt with the production of *alumina*. The text in this exercise goes on to explain how alumina is refined into *aluminium*.

New Vocabulary: *cryolite* *fluorine*

KEY

1 converting 2 electric 3 separates 4 sinks 5 solidify 6 are 7 transformed 8 eventually 9 furnaces 10 lining 11 blocks 12 combination 13 point 14 affected 15 into 16 than 17 bottom

Exercise 8 (Oral)

KEY

1 dried 2 powdered/heated 3 pressurized *or* heated 4 dissolved 5 undissolved 6 filtered 7 crystallized 8 purified 9 liquefied 10 oxidized

SECTION C: The Production of Iron

Teaching Points

1 **Simple Passive for Processes** (cont.)
The use of the Passive for describing industrial processes is further developed in this final Section. Several more irregular verbs are introduced, as well as the *Passive* + by + *Agent* construction (see Exercise 11).

2 **Verb** + *to* + **Infinitive**
Students have recently encountered:
allowed to (sink)
left to (cool)
The Reading Text in this Section also includes:
used to (make cement)
causes (it) *to* (separate)
This construction is dealt with more fully in Unit 7.

The Listening Text

Ores are generally concentrated *by* a combination of different processes. Then the minerals are purified into pure metals by a process called *reduction*. In a reduction process, the pure metal is separated from the other elements in the mineral. In chemical terminology, we say the mineral is *reduced* to a pure metal. Copper sulphide is reduced to 98% pure copper in a converter. Alumina is reduced to pure aluminium in an electric furnace. And now, this Reading Text shows how iron oxide is reduced to pure iron in a *blast furnace*.

In a reduction process, other materials are normally added to the mineral concentrate. These added materials combine with the mineral compound and they *cause* the pure metal *to* separate from the other elements or compounds. For example, in copper production, oxygen was added to copper sulphide minerals. The oxygen separated the copper from the sulphur. Copper sulphide plus oxygen became copper plus sulphur dioxide.

In iron production, the added materials are *coke* and *lime*. The blast furnace's *fed* with a mixture of iron ore, coke and lime. This mixture's called the *charge*. Inside the furnace, the charge's separated into pure iron, carbon dioxide gas, and liquid calcium silicate. The liquid calcium silicate's called *slag*. The slag's easily removed from the liquid iron because it has a different density.

After Listening to the Text

If the students have understood the Listening Text, they should be able to explain for you the following:

a reduction process
a charge
slag

Exercises

Exercise 9 (Oral/Written)

Ask students these questions orally first. They need only give short answers, simply so that their comprehension can be checked. Then they can go on to write out their answers in full.

KEY (short answers)

1 Around 1800.
2 Coal or charcoal.
3 The charge contains coke instead of coal, and hot air is used instead of cold air.
4 Yes, electric furnaces.
5 Iron oxides.
6 It is crushed, powdered and mixed with coke dust and lime.
7 Iron ore, coke dust and lime.
8 A blast of hot air.

9 $FeO + CO \rightarrow Fe + CO_2$.
10 Silica impurities are transformed into calcium silicate.
11 It is less dense.
12 It is either cast into bars or converted into steel.

Exercise 10 (Oral)

Students may look up these irregular verbs in the Appendix.

KEY

1 was burnt 2 are found 3 are broken 4 is ground
5 is fed 6 is blown 7 is left 8 is cast 9 is taken

Exercise 11 (Written)

KEY (type A sentences)

1 A crusher breaks the rock into smaller pieces.
2 A filter removes large unbroken pieces.
3 A blast of hot air oxidizes the coke.
4 Heated carbon monoxide reduces the iron oxide to pure iron.
5 The lime transforms the silica impurities to calcium silicate.
6 The heat of the furnace melts the iron and the slag.
7 A large blast furnace produces 3000 tonnes of iron per day.
8 A separate furnace converts iron into steel.

Exercise 12 (Reading/Vocabulary)

New Vocabulary: *a skip* *see*

This text describes in more detail how the charge is fed into the top of a blast furnace. Encourage students to study the diagrams closely along with the text.

KEY

1 caused 2 exhaust 3 top 4 under 5 allowed
6 by 7 transferred 8 into 9 fall 10 sealed 11 charge
12 never 13 through

UNIT FIVE

Graphs

SECTION A: Mathematical Equations

Teaching Points

1 **Subordinators** (*if* . . . . , (*then*) . . . .)

The complex sentence in English has already been presented to your students with the use of *because* (Book 1, Unit 6). This Unit is mainly concerned with introducing and practising complex sentences (with two clauses only) which include the common subordinators *if*, *as*, *although*, *unless* and *when*. In Section A, only *if* . . . , (*then*) . . . . is presented.

If . . . . , (*then*) . . . .

If has various related usages. In this Unit, we only offer the most simple usage of *if* as a subordinator to introduce sentences containing two interdependent propositions (i.e. no 'conditionals' are involved).

EXAMPLES: *If* x = 1, (*then*) y = 1.
If the coefficient of x is high, (*then*) the gradient is high.

NOTE: The use of *then* in the second clause is optional. When first introducing *if* sentences to students, the *then* can be included – if for no other purpose than to 'mark' the beginning of the second clause. All the *if* sentences in Section A begin with the *if* clause, although it is, of course, entirely possible to reverse the order of the two clauses. When the subordinate (*if*) clause comes first, it is necessary to put a comma between the two clauses.

2 **Mathematical Terminology**

It is assumed that teachers and students know how to read and make simple graphs. If you are not sure of your students' abilities in this direction, spend a longer time on the **Before Listening to the Text** section below.

Here are some of the mathematical terms used:

Axis: The horizontal/vertical (*x*/*y*) line on a graph, marked off with a scale of measurements. The two *axes* are the reference lines for reading the graph.

Value: In mathematical language, a number or quantity.

Constant: A constant always has the same value (see The Listening Text).
Variable: *x* and *y* are variables, i.e. their values can vary (see The Listening Text).
Origin: The point on the graph where the *x* and *y* axes cross (see The Listening Text).
Gradient: The angle between the graph line and the *x* axis. A gradient can be 'high' (steep) or 'low' (gradual) (see Reading Text). The gradient of a graph can be given a numerical value (see Exercise 4).
Coefficient: This is a difficult term. It suffices to know, at this stage, that in the expression *2x*, the coefficient of *x* is *2*. If the variable stands alone, with no number before it, the coefficient is 1, i.e. *x* is really 'one x'.

Before Listening to the Text

Sketch a grid (like an empty crossword puzzle) on the board. Mark the points along the left-hand side and base from 0–10, starting from the bottom left-hand corner. Put a *y* at the top of the vertical numbers and an *x* at the end of the horizontal numbers. Introduce the phrases *graph* and *x/y axis/axes* (see LANGUAGE NOTE 13).
Ask your students to come up to the board and mark points on the graph as you call them out.

TEACHER:	x = 1, y = 2. (etc.)	STUDENT:	Marks appropriate point on the graph.

Try and call out a set of points that will form a line (either curved or straight). When the points are all marked correctly, draw in the graph line, and ask students to read off the values of *x* and *y* at any points you care to choose.

The Listening Text

Mathematics uses many technical terms. For example, there's the word *constant*. A constant always has the same *value*. Its values don't change. π (*pi*) is a *mathematical* constant. $\pi = \frac{22}{7}$ (twenty-two over seven). It has a value of 3·143. The speed of light is also a constant. The speed of light doesn't vary.

However, most measurements aren't constants. They're *variables*. That's to say, they don't have a constant value. Temperature's normally a variable and so's pressure. *Even* the boiling point of a liquid's a variable. (*Refer to the graph of the boiling point of water, Book 1, p. 69, if you wish.*) Two different variables are often *related*. In our example, the b.p. of water's related to atmospheric pressure. *If* the pressure's normal, *then* the b.p.'s 100°C. If the pressure's higher, then the b.p.'s above 100°C. If the pressure's below normal, then the b.p.'s below 100°C.

(*Now refer to the illustrations in the present section.*) A graph shows the *relationship* of two variables. The variables are generally labelled x and y. The values of y are shown along the vertical axis and the values of x along the horizontal axis. The two axes cut at a point. At this point, the values of the two variables are often zero. The point's called the *origin* of the graph.

The relationship of two variables is also given in the form of an equation. The graph line's calculated from this equation and the equation's generally *written* along the graph line.

After Listening to the Text

Repeat and drill the three *if* sentences in paragraph two above. Then use any graph to practise:

If $x = \ldots.$, *(then)* $y = \ldots.$

Finally, use the question pattern:

If $x = \ldots.$, *(then) what does y equal?*

Exercises

Exercise 1 (Reading/Written)

The graph equations are all illustrated above the text. Those statements which are false (except one) can be corrected by putting a negative into the original sentence. The exception is the last sentence (6e) where the sentence is correct for two of the equations (x^2, x^4), but but not for the other (x^3).

KEY

1 (a) True.	(b) True.	(c) False.	
2 (a) True.	(b) False.	(c) False.	(d) True.
3 (a) False ($y \neq 0$).	(b) True.	(c) True.	(d) False.
4 (a) True.	(b) True.	(c) False.	
5 (a) False.	(b) True.	(c) False.	(d) False.
6 (a) True.	(b) True.	(c) True.	(d) False (for x^3).

Exercise 2 (Oral/Reading)

KEY (short answers)

1 yes   2 yes   3 $y = -20$ 4 twice   5 $y = 2x^2 - 20$

6 -20 7 2   8 *If* $x = 1, y = -18$

$= -2, y = -12$

$= 4, y = 12$

$= 5, y = 30$

Exercise 3 (Written)

KEY

1 If the constant is positive, the line passes above the origin.
2 If the constant is negative, the line passes below the origin.
3 If x has a high coefficient, the gradient is high.
4 If x has a low coefficient, the gradient is low.
5 If the equation contains x^2, the graph (line) is curved.
6 If x is negative, x^4 is (always) positive.
7 If angles A, B, C are right angles, angle D is (also) a right angle.
8 If angles A and B are 45°, angle C is a right angle (90°).
9 If angle B = 90° and angle C = 60°, angle A = 30°.
10 If m = 500 g and V = 100 cm^3, $\rho = 5\ g/cm^3$.
11 If $\rho = 2\ g/cm^3$ and V = 16 cm^3, m = 32 g.
12 If $\rho = 9\ g/cm^3$ and m = 900 g, V = 100 cm^3.

Exercise 4 (Written/Reading)

The report can be written in the same way as that described for laboratory reports in Unit 3. However, no *Equipment* heading is needed.
The gradients for the other three lines are:

(line x) $\frac{QR}{PR} = \frac{2}{1} = 2$

(line y) $\frac{1}{1} = 1$

(line z) $\frac{0 \cdot 5}{2} = 0 \cdot 25$

SECTION B: Boyle's Law

Teaching Points

1 Subordinators (cont.)
This Section introduces three more subordinators:

as . . . . , (*so*) . . . .
although . . . .
unless . . . .

A *as* . . . . , (*so*) . . . .
as has several different usages in English. The one used here is:

as the pressure increases, (*so*) the volume decreases.

In *as* . . . . , (*so*) . . . . , the *so* is optional in the same way as *then* in *if* . . . . , (*then*) . . . .

B *although*
This word can often be used to make *either* of the two clauses in a sentence subordinate to the other (see Exercise 6), but a shift in meaning will result.

EXAMPLES: *Although* Graph A gives some information, it is incomplete.
Graph A gives some information *although* it is incomplete.

C *unless*
Some textbooks assume that *unless* is an equivalent of *if* . . . . *not*. This is a dangerous assumption and one we have avoided making.

In Section B, the subordinators appear both in the sentence-initial and sentence-medial positions. Point out that if the subordinate clause is after the main clause, there is usually no comma separating the two.

2 Days of the Week
Exercise 8 is devoted to the introduction and practice of the days of the week in English and their abbreviations.

The Listening Text

There are normally two variables in a graph or an equation. For example, let's look at the composition of steel. (*Refer to the graph in Book 1, p. 105, if you wish.*) The hardness of steel's related to its carbon content. *As* the carbon content increases, *so* the hardness increases. Similarly, as the carbon content increases, so the strength of steel also increases.

One of the two variables is called the *dependent* variable and the other's called the independent variable. In this example, hardness (or strength) *depends on* carbon content. Carbon content doesn't depend on hardness or strength. Therefore hardness is the dependent variable and carbon content's the independent variable.
Although there are two variables, one is dependent on the other. The dependent variable's always shown along the y axis, and the independent variable's shown along the x axis. The dependent variable's always written on the left in the equation. Although $2x = y$ is the same as $y = 2x$, it's normal to write it with the y on the left.
It's very important to label a graph correctly. For example, *unless* the two axes are labelled, it's difficult to extract information from the graph. A graph isn't complete unless it's labelled correctly.

Exercises

Exercise 5 (Oral/Reading)

KEY

A **Boyle's Law**
The pressure of a gas varies in inverse proportion to its volume.
As the pressure increases, (so) the volume decreases.
Conversely, as the pressure decreases, (so) the volume increases.
Charles' Law
The volume of a gas varies in direct proportion to its temperature.
In other words, the volume decreases as the temperature decreases.
Conversely, as the temperature increases, the volume increases.

B 1 units of measurement 2 graph equation
3 the values of p and v 4 'Temperature Constant'

Exercise 6 (Oral/Written)

KEY

1–5 *although* can be placed in sentence-initial or sentence-medial position (see Teaching Points).
6 unless 7 as 8 because 9 unless 10 as

Exercise 7 (Reading/Vocabulary)

KEY

1 relationship 2 Law 3 direct 4 words 5 increases
6 conversely 7 decreases 8 negative 9 although
10 impossible 11 marked 12 to 13 above

Exercise 8 (Listening)

Remind students that the days of the week begin with a capital letter in English. After presenting and practising the seven days orally, ask students to fill in the first table, ensuring that their spelling is correct.

Monday (*Mon.*) *Tuesday* (*Tues.*) *Wednesday* (*Wed.*)
Thursday (*Thurs.*) *Friday* (*Fri.*) *Saturday* (*Sat.*)
Sunday (*Sun.*)

NOTE: In European calendars, *Sunday* may be the first or the last day of the week.

The Listening Text (Pause = //)

Here is a different kind of graph. It's a graph of the maximum temperatures for each day of a week. Complete the graph as you hear the information. First, write MAXIMUM DAILY TEMPERATURES FOR THE WEEK above the graph. //

Next, label the axes. Label the horizontal axis DAYS OF THE WEEK // Put the abbreviation for each day below each mark along the axis. Start with MON. on the left and finish with SUN. on the right. //

Now label the vertical axis. Label it '°C' and mark it from 18 at the bottom of the axis to 26 at the top. //

Now listen carefully. On Monday, the maximum temperature was 20°. Mark the point on the graph. // Now mark the correct points for the other six days in the same way.

On Tuesday, the maximum temperature was 23°. //
On Wednesday, it was 19°. //
On Thursday, ... 25°. //
On Friday, ... 25° again. //
On Saturday, ... 24°. //
On Sunday, ... 18°. //

Now join each point on the graph to the following one. This gives you the graph line. // Look at the line. It's not a straight line and it's not curved either. It's called a broken line.

SECTION C: Time Graphs

Teaching Points

1 Subordinators (cont.)
When . . . . is introduced in the Reading Text. This subordinator may cause some difficulties as it is sometimes interchangeable – though not really synonomous – with *If* . . . . (see Exercise 11).

2 Some Prepositional Usages
The students' attention is drawn to several important uses of common prepositions in this Section.

A *by/to* are contrasted in sentences like:

The voltage decreased (from 1·4 V) *to* 1·3 V.
It decreased *by* 0·1 V.
(See Reading Text and Exercise 10.)

B *in/for*. Both these prepositions are used to signal a duration of time.

EXAMPLES: The voltage decreased by 0·2 V *in* the last two hours.
The cell became useless when it was used *for* more than eight hours.

C *in/on*. The days of the week (with *on*) were introduced in Section B. The months of the year (with *in*) are introduced in Exercise 12 (see also LANGUAGE NOTE 17).

3 *Any*
Two different uses of *any* are presented, one in The Listening Text and the other in the Reading Text.

A *any* in the Listening Text is used thus:

. . . to measure the voltage at *any* point in time.
. . . at *any* other hour of the day.

B In the Reading Text, it is used differently:

. . . do not give *any* information about. . . .
. . . difficult to take *any* accurate measurements.

4 Months of the Year
Exercise 12 is devoted to the presentation of the months of the year and their abbreviations in English.

Before Listening to the Text

A presentation (or revision) of the units of time measurement is appropriate at this point. First, elicit and practise the units that students already know:

minute, *hour*, *day*, *week*, *year*

Then introduce the two new units used in this Section:

second, *month*

Give oral practice:

TEACHER
or
STUDENT — How many . . . . in a . . . . ? (hours/day etc.)

STUDENT — There are . . . . in a . . . . (24 hours/day)

Next, revise year dates (introduced in Unit 2). Use sentences including dates such as:

Parsons' first turbine was built *in 1884*.

Draw attention to the use of *in*. Contrast it with the use of *on* in sentences like:

There is an English lesson *on Monday*.

The Listening Text

Time's often one of the variables in a graph. It's always the independent variable. If you look at a time graph, you always see the time scale along the horizontal axis.

Time graphs are either *continuous* or *discontinuous* graphs. Continuous graphs have a curved or straight graph line. A discontinuous graph has a broken line. Look at the graphs in this Section. The two above the Reading Text are continuous. It's possible to measure the voltage at *any* point along the line. The graphs in Exercise 10 are also continuous, but the graph in Exercise 12's discontinuous. It contains two broken graph lines. It's impossible to take any measurements between points on the line.

You *made* a discontinuous graph in Exercise 8. The graph line was broken. Although the graph measured temperatures, it only measured the temperature at one point per day. It only *gave* the maximum temperature for each day. It gave no information about any other temperature in the day. A discontinuous graph only gives accurate information at fixed points along the graph line. It's impossible to take any measurements between these fixed points.

After Listening to the Text

Ask students to look at graph P before they start reading. Make sure they can read it correctly by asking 'comprehension' questions on it.
Finish off by giving oral practice with the following pattern:

In the first (two) hour(s), the voltage decreased from . . . .
next to . . . .
last by . . . .

Exercises

Exercise 9 (Written/Reading)

KEY

1 True.
2 False. The voltage decreases.
3 True.
4 True.
5 True.
6 False. The voltage scale is suitable.
7 True.
8 False.
9 False.
10 False.

Exercise 10 (Written)

It may be advisable to go through the exercise orally first and obtain short answers from the students, in order to check their reading of the graphs.

KEY (short answers)

1 From 1·2 V to 1·1 V. . . . by 0·1 V in 2 hours.
2 From 1·1 V to 1·0 V. . . . by 0·1 V in 2 hours.
3 From 0 km/h to 40 km/h. . . . by 40 km/h in 5 seconds.
4 From 40 km/h to 70 km/h. . . . by 30 km/h in 5 seconds.
5 From 70 km/h to 95 km/h. . . . by 25 km/h in 5 seconds.
6 From 10°C to 70°C. . . . by 60°C in 10 minutes.
7 From 70°C to 100°C. . . . by 30°C in 10 minutes.
8 From 100°C to 90°C. . . . by 10°C in 10 minutes.

Exercise 11 (Oral)

KEY

1 when	2 when	3 when *or* if	4 when *or* if
5 when *or* if	6 when	7 if	8 when *or* if

Exercise 12 (Listening)

New Vocabulary: *average* *rise* *stay* *a difference*
how long . . . . ? *how far . . . . ?*

Introduce and practise the months of the year and their abbreviations in the same way as you did for the days of the week in the last Section. Students should fill in the first table after oral practice:

January (*Jan.*) *February* (*Feb.*) *March* (*Mar.*) *April* (*Apr.*)
May *June* (*Jun.*) *July* (*Jul.*) *August* (*Aug.*) *September* (*Sept.*)
October (*Oct.*) *November* (*Nov.*) *December* (*Dec.*)

The Listening Text

Here's another discontinuous time graph. Time's the independent variable and so the time scale's along the bottom of the graph. The vertical axis gives a scale for temperature from −3°C to 24°C.

There are two graph lines. The graph gives information on the *average* monthly temperatures in both New York and London. The dotted line's for New York. New York has both higher and lower average temperatures than London – higher temperatures from April to October, and lower temperatures from December to March. In other words, the temperatures in New York are more extreme than the temperatures in London.

In January, the average temperature in New York's below 0°C. In London, the January average's about 4°C. So there's a temperature *difference* of 4°–5° in January. The monthly averages *stay* about the same in February, and then they *rise* in March. The New York temperature rises above the London temperature. July's the hottest month in both New York and London. There's an average temperature difference of about 5°–6° in July. In August, both temperatures fall again. In December, the New York temperature finally falls below the London temperature again.

When you are sure students understand the graph, use the question patterns below it for oral practice. Since the patterns are set out in the Students' Book, they may be asked to question as well as answer.

UNIT SIX | **Motor Vehicle Maintenance**

SECTION A: Lubrication

Teaching Points

1 Modal Verbs

This Unit introduces three of the most common modal verbs – *will*, *may* and *can*. Section A only deals with *will*. A few general points about modals and their use are worth remembering and pointing out to your students as the need arises.

(i) The common modals are each used in various contexts. The 'meaning' of one modal often overlaps the 'meaning' of another.

EXAMPLES: *Will* you pass the spanner?
Can you pass the spanner?

Can I go with you?
May I go with you?

(ii) It is probably more advisable to think of *could*, *would*, *might*, etc. as modals in their own right rather than as Past Tense forms of *can*, *will* and *may*. (See the use of *should* in Unit 7.)

(iii) Modal verbs have no infinitive or Passive forms. They are followed (with some exceptions) by an infinitive without *to*.

(iv) Modals have no *-s* suffix in the 3rd person singular form. They do not take an auxiliary – i.e. questions are formed by inversion without *do*/*did* and negatives are also formed without *do*/*did*.

***Will* and Conditionals**

The most common use of *will* in technical language is for 'prediction'. It is difficult to talk of *will* as being used to mark the 'future' since sentences such as:

(a) Friction *will cause* (causes) heat

is a more common type of 'prediction' than

(b) Cars in the 1980s *will need* no greasing.

In other words, technical English contains more 'timeless' predictions than 'future' predictions. 'Timeless' predictions, such as (a) above may be phrased with *will* or the Present Simple Tense. A third type of prediction, where *will* is almost always used instead of the Present Simple, could be called 'conditional' predictions:

> (c) If there *is* not enough oil, friction *will* increase.
> If the oil *is* not changed, the filter *will* become blocked.

Students should notice that, in type (c) sentences, the 'condition' is phrased in the Present Simple, whereas the 'prediction' is phrased with *will.*

2 Complex Noun Phrases

It is characteristic of technical language that noun phrases tend to be more complex than in everyday language (whereas verb phrases are often less complex). Students have already encountered:

> noun+noun
> noun+'*s*+noun
> noun+*of*+noun

From this point on in the course, we have allowed noun phrases to become more complex. In this Section, we have such phrases as:

> only small particles of dirt or metal
> friction between the moving parts
> cars in the future

3 Complex Verb Phrases

The introduction of modals obviously adds to the complexity of verb phrases. Moreover, this Section introduces the verb + ... *ing* construction with the verb *need*.

Another verb construction, *become* + ... *ed*, is also used. In this phrase, *become* replaces *is/are* and it is therefore really an alternative Passive construction, denoting 'prediction':

> The filter *is* blocked (*already*).
> The filter *becomes* blocked (*after a period of time*).

Since *become* is common in 'predictions', it is often used in the infinitive form after *will*:

> The engine *will become* damaged.

4 *Once/Twice a . . . . Once/Twice every . . . .*

Twice, *three times*, etc. were introduced in Book 1, Unit 8, in the construction *twice as . . . . as*. (The word *once* is a new item of vocabulary in this Section.)

This Section now introduces a new construction:

> *Once a* month, *twice a* week, etc.
> *Once every* three days, *twice every* three hours.

Point out that *a* is used before a singular noun and *every* before plurals (see Exercise 3).

Before Listening to the Text

Introduce the word (and concept) *lubrication*. This is most easily done by a simple demonstration – for example, rubbing your hands together and drawing attention to the friction and heat caused. Explain that if your hands are wet and soapy, they are 'lubricated', i.e. it is not only easier to rub them together, but heat (and wear!) are minimized.
Oil is used for the same purpose of lubrication between the moving parts of an engine.
You can now leaf through the various mechanical drawings in this Unit, deciding which of the labelled parts are moving parts (or contain moving parts). Many of the terms met will not be new to the students, but others will and should therefore be practised orally as they are encountered.
Draw up two lists, one of (moving) parts which need lubrication and one of those parts that do not.
Finish up with oral practice on the following patterns:

A . . . . needs lubricating, but a . . . . doesn't.
A . . . . will need lubricating, but a . . . . won't.

NOTE: The irregular form *won't* will need pointing out (see LANGUAGE NOTE 18).

The Listening Text

(*The first two paragraphs should be used twice – once as they stand, and then with* will *incorporated into each verb phrase.*)

Combustion causes a large amount of heat inside an engine, and so any petrol or diesel engine *needs* cool*ing*. Most engines have water-cooled engines. The water *flows* around inside the engine. It extracts *some of* the heat and disperses it into the atmosphere.

Friction also causes heat in an engine. All the *moving parts* need *lubricating* with oil. Without constant *lubrication*, they quickly become too hot. The hot metal expands, and the bearings and joints become *damaged*. The oil flows around the engine at high pressure. It covers all the moving parts and reduces friction to a minimum.

No vehicle runs well without the correct amounts of oil, *grease* and water. That's why it's very important to check the oil and water supply *regularly*. They will need checking *once every* two or three days, or at least *once a* week. However, most vehicles *won't* need greasing more than once every six months.

Exercises

Exercise 1 (Reading/Written)

KEY

1 True.
2 False. . . . it is used for cleaning the oil.
3 True.
4 False. Some of the oil will leak out.
5 True.
6 False. . . . it is used for checking the amount of oil.
7 False. . . . it will also need changing regularly.
8 True.
9 True.
10 False. . . . they will probably need no greasing.

Exercise 2 (Oral)

In No. 4, make sure students use the oral form *won't*.

Exercise 3 (Oral/Written)

KEY

1 once a week 2 once a day 3 once every 3 months
4 once every 15 minutes 5 once every 3 seconds 6 once a second
7 once every 2 weeks

Exercise 4 (Reading/Vocabulary)

New Vocabulary: *anti-freeze*

KEY

1 causes	2 will	3 needs	4 cooled
5 air	6 stroke	7 flows	8 reduces
9 radiator	10 out of	11 leak	12 boil
13 amount	14 checking	15 below	16 unless

SECTION B: Filters

Teaching Points

1 **Modal Verbs** (cont.)
The second modal verb to be introduced is *may*. The most common use of *may* in technical language, and the one presented here, is to express 'possibility':

It *may* be made of paper or it *may* contain oil.

In this use, it can be contrasted with *will*, used to predict with more certainty:

The engine *will* (certainly) not operate efficiently,
and damage *may* (perhaps) be caused (see Exercise 6).

2 *Be*
The infinitive form of *is/are* has not yet been encountered by students. Here, it is used in the Passive Infinitive (after modals). *Be* + ... *-ed* is practised and contrasted with *become* + ... *-ed* in Exercise 8.

Before Listening to the Text

Students have previously been introduced to the common oral question patterns:

What is . . . . for?
and *What does . . . . do?*

They have also already had the opportunity to look through and discuss the parts labelled in all the drawings in this Unit. Leaf through the drawings again (especially the two large drawings at the end of this Section and the next), using the above question patterns to ask about the labelled parts. Take care which question pattern you use on which occasion. For example:

	What does a carburettor do?
	(It mixes the petrol with (filtered) air.)
is better than:	What's the carburettor for?
whereas:	What's the gear lever for?
	(It's for changing the gears.)
is better than:	What does the gear lever do?

The Listening Text

A petrol engine needs oil and water. It also needs petrol and air. The petrol and the air are mixed inside the *carburettor*. The petrol and air mixture passes from the carburettor into the top of the cylinders through a valve. The mixture's then exploded by the spark plug. The combustion forces the piston down the cylinder.

(*To contrast* will *and* may)
The engine won't operate well if the petrol or the air aren't filtered *before* they go into the cylinder. Any dirt or other impurities will decrease the engine efficiency and they *may* cause damage. Therefore any vehicle will have both a petrol and an air filter. There may be two petrol filters, one inside the carburettor and the other in the fuel pump. There will be only one air filter, but it may be one of several different types. The petrol filter may not need cleaning for several years, but the air filter will need cleaning or replacing regularly.

Exercises

Exercise 5 (Oral/Reading)

These questions can be rephrased with *may* or *will*. Similarly, students can be asked to answer using *may* or *will* as appropriate.

Exercise 6 (Oral)

KEY

1 may 2 will 3 will 4 will 5 will 6 may. . . . may

Exercise 7 (Reading/Vocabulary)

KEY

1 fuel	2 from	3 passes	4 particles
5 impurities	6 cause	7 needs	8 blocked
9 becomes	10 stop	11 between	12 contains
13 pump	14 collect	15 remove	16 clean

Exercise 8 (Written)

KEY

1 The pistons and the piston rings will be damaged by dirt in the cylinders.
2 The spark plugs may become covered with oil from the cylinders.
3 The amount of fuel in the mixture will be increased by a blocked air filter.
4 Considerable damage to an engine may be caused by too much fuel in the mixture.

5 Friction between moving parts will be decreased by lubrication.
6 The density of a liquid will be changed by impurities.

(Oral)

7 becomes . . . . be
8 becomes . . . . be
9 be . . . . become
10 be . . . . becomes

SECTION C: Servicing

Teaching Points

1 Modal Verbs (cont.)

Can, like *may*, chiefly expresses 'possibility'. The predominant meanings of these two modals overlap considerably.

EXAMPLES: If vehicles are not serviced, faults *can* occur.
If vehicles are not serviced, faults *may* occur.

In other cases, it is possible to detect a shift in meaning between the two.

EXAMPLES: . . . damage *may* be caused (*perhaps*).
. . . damage *can* be caused (*a physical possibility*).

This emphasis on 'physical possibility', when *can* is used, is even more evident when the verb is in the negative:

The gap *can* be adjusted with a screwdriver, but the points *cannot* be cleaned.

This more distinct use of *can–cannot* is brought out in Exercise 10. Draw attention to the negatives of *can–cannot* (one word) in written language and *can't* (where the vowel is lengthened) in spoken language (see LANGUAGE NOTES 20 and 21).

The Listening Text

A car engine's a very complex device. If it isn't regularly *serviced*, it *can* easily become inefficient. Most of the moving parts are affected by wear and need regular servicing or replacing.

(*Use the diagram for Exercise 11.*)
A four-cylinder car, for example, has eight valves – two for each cylinder. Each of these valves opens and closes several thousand times a minute, so the valves can easily become *worn* or broken. The valve springs will also become less elastic as they are used. As the valves become worn, the

engine will become less efficient. A worn valve *can't* seal the cylinder correctly. Similarly, a weak valve spring can't operate the valve correctly.

(*Use the diagram above the Reading Text.*)
Not only mechanical parts become worn. The electrical parts also need regular servicing. Each spark plug produces a spark several thousand times a minute. The *rotor arm* in the *distributor* can *rotate* at more than 2000 rpm. The *points* inside the distributor can open and close at four times this speed. The electrodes on the spark plugs will become worn. As they wear, the gap between them will increase and the spark will become weaker. The end of the rotor arm and the points will also become worn from constant use. The spark plug gap can be *adjusted*, but you can't adjust the rotor arm. It can only be replaced when it's worn.

After Listening to the Text (or after reading the text)

Ask students to make a list, using all the diagrams in the Unit, of the things that need checking regularly. The list can be headed: SERVICING or MAINTENANCE. They should also indicate what sort of fault is likely to occur in each case and what needs to be done if a fault is revealed.
A complete list would include:
Engine oil – oil level low (fill up) – oil dirty (change oil and oil filter).
Gearbox and differential – oil level low (fill up).
Water in radiator – water level low (fill up).
Piston rings – worn (replace).
Air filter – dirty (replace or clean depending on type).
Carburettor – dirty (clean).
Fuel filter – dirty (clean).
Fan belt – worn or broken (replace).
Spark plugs – dirty (clean) – worn (adjust or replace).
Rotor arm – worn (replace).
Points – worn (replace or adjust).
Battery – electrolyte level low (fill up with distilled water).
Valves and valve springs – worn (replace).
Rocker arm gap – (adjust).

Exercises

Exercise 9 (Reading/Written)

KEY

1 correct	2 correct	3 incorrect	4 correct
5 incorrect	6 incorrect	7 correct	8 incorrect

Exercise 10 (Oral)

KEY

1 can	2 can't	3 can't	4 can	5 can	6 can't

Exercise 11 (Reading/Vocabulary)

KEY

1 may	2 because	3 efficient	4 each
5 eight	6 worn	7 cannot	8 valves
9 adjusting	10 be	11 check	12 less

Exercise 12 (Oral/Written)

KEY

1 Check the gap between the spark plug electrodes.
2 Check the amount of oil in the sump.
3 Check the gap between the valve and the rocker arm.
4 Check the amount of oil in the gearbox.
5 Check the condition of the valve springs.
6 Check the amount of oil in the differential.
7 Check the amount of electrolyte in the battery.
8 Check the air pressure in the tyres.
9 Check the amount of water in the radiator.

UNIT SEVEN | **Welding**

SECTION A: Gas Welding

Teaching Points

1 Modal Verbs (cont.)
Must normally implies *necessity*:

> In gas welding it is *necessary* to use a mixture of two gases.
>
> A mixture of two gases *must* be used.

Must not and *need not*.
We can say that *must* has two negative forms. *Must not* also implies 'necessity':

> The oxygen *must not* be turned on before the flame is lit. (i.e. It is *necessary* to light the flame before the oxygen is turned on.)

Need not, on the other hand, implies 'no necessity'.

> The combustible gas *need not* be acetylene.
> (i.e. It is not necessary to use acetylene.)

NOTE: *Need not* is a true modal verb and has all the grammatical features of the other modals used so far (described in the Teaching Points of Section A).

2 *To+Infinitive* (Infinitive of Purpose)
Students have already encountered:

> *It is*+adjective+*to*+infinitive
> *Allowed to*+infinitive,
> etc.

However, the Infinitive of Purpose signifies a somewhat different usage of *to*+infinitives. For example, it is possible to begin a sentence with *to*:

> The oxygen supply must be adjusted *to give* the correct flame.

To give the correct flame, the oxygen supply must be adjusted. (See Exercise 3.)

To, as used here, is an equivalent of *in order to*. The full phrase *in order to* is presented in Section B.

3 Verb+*ing*

(i) The verbal form may be used as a noun denoting an action or process (e.g. engineering or soldering). The last Section was called *Servicing* and the present one is called *Welding*.

(ii) Verbal forms of this type are also used adjectivally, as in:

A *welding* torch.
An *oxidizing* flame.
Adjusting valves.

The Listening Text

Pieces of metal can be joined in several ways. They may be bolted or riveted, or they can be joined with solder. In construction work they are often *welded* together.

Soldering and *welding* are different processes. In soldering, the *workpieces* aren't melted. Only the solder's melted. In welding, the workpieces *must* be melted as they are joined. A *welded* joint usually has as much strength as the original workpieces.

In welding, the workpieces must be made of similar metals. They *needn't* always have exactly the same composition, but they must be similar. For example, you can weld low carbon steel to medium carbon steel, but you can't weld a piece of tin to a piece of steel. When a *filler metal*'s used, it must also be made of a similar metal. The filler metal's not used to join the workpieces. It's only used to fill the gaps between them.

In gas welding, a mixture of two gases is used *to provide* the *flame*. When oxygen's added to a *combustible* gas, the mixture will burn at a very high temperature. The heat of the *welding* flame's increased by the oxygen. That's why oxygen must be used. Without it, the flame's not hot enough to melt the workpieces.

Oxygen cylinders are normally painted black and *acetylene* ones are red. Oxygen cylinders are also longer and thinner than acetylene ones. Here is a *caution*: Acetylene *mustn't* be stored at high pressure. It's very dangerous.

After Listening to the Text

Present and practice orally *must/mustn't/needn't*. Here are some examples that can be used. Use the first one for presentation and others for practice.

EXAMPLES: *Gas Welding*

A combustible gas *must* be used in gas welding.
Acetylene *mustn't* be stored at high pressure.
The combustible gas *needn't* be acetylene.

Battery Maintenance

Distilled water *must* be used to refill battery cells.
Tap water *mustn't* be used.
The cells *needn't* be filled completely.

To Calculate the Density of a Metal

The balance *must* be accurate.
The block *mustn't* be dirty.
The block *needn't* be regular in shape.

Exercises

Exercise 1 (Reading/Written)

KEY

1 False. . . . it need not be used.
2 False. . . . oxygen must be mixed.
3 True.
4 False. . . . acetylene must not be stored at high pressure. (i.e. it *can* be stored at high pressure, but it is dangerous.)
5 False. . . . . low gas pressures.
6 True.
7 True.
8 False. . . . bronze or brass.
9 True.
10 False. . . . the oxygen supply.

Exercise 2 (Written)

KEY

1 . . . two gases must. . . .
2 . . . a combustible gas must. . . .
3 . . . oxygen must. . . .
4 . . . acetylene need not. . . .
5 . . . pressure regulators must. . . .
6 . . . steel welding rods need not. . . .
7 . . . low pressures must. . . .
8 . . . high pressures must not. . . .

Exercise 3 (Written)

KEY

1 A welding torch . . . . melt the workpieces.
2 Gas welding . . . . join steel to steel.
3 Oxygen . . . . provide a hot enough flame.
4 Pressure regulators . . . . regulate the gas pressures.
5 Adjusting valves . . . . adjust the welding flame.
6 Welding rods . . . . supply filler metal.

7 To adjust the welding flame, the oxygen pressure can be adjusted.
8 To create the welding flame, it is necessary to mix two gases.
9 To compress acetylene, it is dissolved in acetone.
10 To light the welding torch, the combustible gas must be turned on first.
11 To make a very strong joint, the workpieces must be composed of the same metal.

Exercise 4 (Reading)

New Vocabulary: *an excess*

KEY

1 to 2 regulate 3 types 4 created 5 work 6 mixture
7 welding 8 contains 9 welded 10 flame 11 supplied
12 different 13 determines 14 necessary 15 larger

SECTION B: Arc Welding

Teaching Points

1 *In order to* and *or else*

A (*in order*) *to*

The full phrase used with the Infinitive of Purpose construction is *in order to*. The first two words of the phrase, as we have seen in the last Section, are optional. The full phrase is not as common as the plain infinitival form (. . . . *to* . . . .), especially in speech. The phrase is practised in Exercise 7.

B *or* (*else*)

The full phrase can be shortened to *or* (in written or spoken language). It is advisable to introduce it first, as we have done, in the full form to avoid confusion with the more elementary use of *or* as a simple conjunction. *Or* (*else*) is practised in Exercises 6 and 7.

2 **Sign Language**

For reasons of practical safety, one most important area of language for students to comprehend is that of written signs which act as warnings, cautions or operating instructions. Some elementary sign language is presented in Exercise 8.

The Listening Text

Gas welding can be quite a slow process. A welding torch doesn't produce temperatures above about 2500°C. Under normal conditions, the average welding speed for steel plates, with gas welding, is only about 4 metres an hour.

Arc welding's much quicker. The welding *arc* can produce much higher temperatures than 2500°C. This arc's supplied by a strong electric current. Look at the diagram above the Reading Text. The electric current from the welding equipment creates an arc between the electrode and the workpiece. The temperature of this electric arc is normally 4500°C or more.

A powerful electric current must be provided, *or else* the arc won't have a high enough temperature. The normal mains supply isn't suitable for arc welding equipment. It doesn't have enough *amperage*. However, mains *voltage* is quite high. So a *transformer*'s used *in order* to lower the voltage and raise the amperage. The voltage must be lowered in order to make the equipment less dangerous to operate. The amperage must be increased in order to supply a strong enough current for the arc. So a

transformer must be used or else the equipment won't be either safe or efficient.
The electrode rod's composed of filler metal. As the arc melts the workpieces, it also melts the rod. The melting rod fills any gaps between the workpieces. When it becomes too short, it's replaced with a new rod.

Exercises

Exercise 5 (Oral/Written)

KEY (short answers)

1 By the heat of the arc.
2 By a strong electric current.
3 No. . . . only 1 A.
4 To prevent overheating . . . . because the current is strong.
5 To protect the operator (from electrocution).
6 To provide the filler metal to join the workpieces.
7 To supply the necessary current.
8 To complete the electric circuit.
9 A (secondary) arc will appear between the clamp and the workpiece.
10 (i) Switch on the transformer.
(ii) Brush the electrode against the workpiece.
(iii) Withdraw the electrode from the workpiece (not too far).
(iv) Hold the electrode end about 4 mm above the workpiece.
(v) Move the electrode across the joint continuously.

Exercise 6 (Written)

KEY

1 The current must be above 60 A . . . . the arc will not be strong enough.
2 The cables must be thick . . . . they will overheat.
3 The electrode must not be moved too quickly . . . . the electrode will not melt.
4 An earth clamp must be attached to the workpiece . . . . the current will not flow.
5 The clamp must be attached securely . . . . arcs will appear in the wrong place.
6 The welding current must be at a low voltage . . . . there is a danger of electrocution.
7 The electrode must not be withdrawn too far . . . . the arc will disappear.
8 The electrode must be moved across the joint continuously . . . . it will become attached to the workpiece.

Exercise 7 (Written)

KEY

1 The power must be switched on in order to create the arc.
2 The current must be at least 60 A in order to create a strong enough arc.
3 The cables must be well insulated in order to protect the operator.
4 The transformer must be connected to the mains in order to provide the necessary current.
5 The electrode must be moved across the joint continuously or else it will become attached to the workpiece.
6 The electrode must not be withdrawn too far or else the arc will disappear.
7 The earth clamp must be securely attached or else an arc will appear between the clamp and the workpiece.
8 The electrode holder must contain an electrode rod or else the workpieces cannot be joined.

Exercise 8 (Oral/Vocabulary)

New Vocabulary: *a sign* *poisonous* *handle* *danger* *poison*

These are only a few of the possible warnings, etc. students may need to be aware of in their working and non-working hours. They can be encouraged to think up other situations where a sign or notice would be needed, and to attempt writing such a warning.
Many more notices of the DO NOT . . . . variety exist – DO NOT TOUCH is a very common one. For further examples, concentrate on machine-operating instructions, freight-handling signs, and danger warnings on electrical equipment.

SECTION C:
Safety Precautions

Teaching Points

1 **Modal Verbs** (cont.)
Should is the last modal to be introduced in Book 2. In its main usage, and the one exemplified here, it is very close in meaning to *must*.

EXAMPLE: You *must* wear goggles to protect your eyes.
You *should* wear goggles to protect your eyes.

However, the two modals can be distinguished. Only *must* can be used in sentences implying 'absolute necessity'.

EXAMPLE: In gas welding, two gases *must* be used.
(i.e. it is absolutely necessary)

In contexts where the necessity is not absolute, but 'advisable' (as in Safety Precautions), *must* tends to be a stronger 'imperative' than *should*.

EXAMPLE: You *must* wear goggles.
(i.e. without goggles, serious injury is highly likely)
Gas cylinders *should* be in a vertical position
(or else an accident may occur).

This distinction is even clearer in negative sentences.

EXAMPLE: You *mustn't* use water on electrical fires.
You *shouldn't* be late for school.

NOTE: These distinctions can be explained to students. However, the choice between *must* and *should* in a particular sentence is often a matter of personal choice. For this reason, we have not emphasized any contrast between the two modals in exercises. However, in Exercise 11, the distinctions may be brought out if desired by discussing which of the eight instructions are more 'imperative' than the others and by using *must* or *should* accordingly.

2 . . . . *so that* . . . .
In its main usage, this subordinator can be paraphrased as . . . . *with the result that*. . . . It is more common in written English where it is often used as an alternative construction to the Infinitive of Purpose.

EXAMPLE: Long sleeves should be worn *to* protect your arms.
Long sleeves should be worn *so that* your arms are protected.
(See Exercises 9 and 10.)

3 Parts of the Body and Names of Clothes
This area of vocabulary is always important. It is made more important by the fact that the same words are commonly used to refer to many specific parts of machines, etc. (See Exercise 12.)

The Listening Text

Arc welding equipment operates with high currents. Currents of two hundred amps or more may be used. If the operator touches the electrode rod or an uninsulated cable, he will get an *electric shock*. He may then become *unconscious*. If this happens, this is what you must do:

1. First, you must turn off the equipment at the mains. You must not touch the operator or the equipment. It is very important that the equipment is turned off first. If the equipment cannot be turned off at the mains, the mains plug must be removed from the socket.

2. If the equipment cannot be turned off and the plug cannot be removed from the socket, the operator *should* be moved away from the equipment. Do not touch the operator. If you touch him, you may receive an electric shock. Move the operator away from the equipment with something wooden or plastic. Never use metal.

3. You should now place the operator on the ground with his arms at his sides. If it is necessary, use *artificial respiration*.

4. Next, you must call for a doctor and an ambulance. Even if the operator is conscious, he must be examined by a doctor.

5. The welding equipment must be carefully examined by an *electrician* before it is used again. The electrician should look for uninsulated wires, *faulty* connections and so on. The equipment must not be used unless it has been checked.

After Listening to the Text

1 Introduce and practise orally *the parts of the body*. The most useful ones for technical students are those shown in Exercise 12. However, you may consider others such as *mouth*, *nose*, *finger*, *foot*, *toe* worth adding to the list. As each new word is introduced, write it on the board. You can check students' retention of these words at the end by saying:

Touch
Move } your . . . .

2 Introduce and practise orally *names of clothes*. Important ones for your students are shown in Exercise 12, but again others such as *sock*, *shoe*, *collar*, *cuff*, *button* may be added as required. To practise the names of materials, ask students what each item of clothing is made of (colours can also be revised, if desired). To practise *clothes* and *parts of the body*

at the same time, ask where each item of clothing is worn (*on the head*, *over the arm*, etc.).

Exercises

Exercise 9 (Oral/Written)

1–4 Oral answers should be in the form:

You must/should wear . . . . to protect . . . .

Written answers should be:

. . . . must/should be worn so that your . . . .

5–8

New Vocabulary: *an electric shock*

Oral answers should be with *to*+verb, and written answers with *so that*. . . .

Exercise 10 (Oral/Written)

1–6 Again, make students give oral answers with *to*+verb, and written answers with *so that*. . . .

KEY (Nos. 7–10)

7	All valves should be turned off	to prevent gas leaks. or gas leaks may be caused.
8	Clean clothes should be worn	to prevent an explosion. or an explosion may be caused.
9	All switches should be turned off	to prevent accidents. or accidents may be caused.
10	The correct precautions should be followed	to prevent injuries. or injuries may be caused.

Exercise 11 (Oral/Written)

Students may decide for themselves whether each sentence should be phrased with *should* or *must*.

KEY

1 long sleeves **2** thick shoes **3** welding gloves **4** a face mask
5 a dry floor **6** tap turned off **7** CO_2 fire extinguisher
8 a metal container

Exercise 12 (Oral/Vocabulary)

New Vocabulary: *a clock* *a needle*

KEY

1 a water jacket **2** a fan belt **3** a (bottle) cap **4** a tooth (of a saw) **5** a head (of a hammer) **6** a piston ring **7** a metal sleeve **8** a (bottle) neck **9** a rotor arm **10** the face (of a speedometer) **11** the hands (of a clock) **12** a table leg **13** an eye (of a needle) **14** the jaws (of pliers) **15** a lip (of a beaker)

UNIT EIGHT

Electrical Instruments

SECTION A: The Instruments in a Car

Teaching Points

1 The Continuous Tenses

In more general, non-technical English courses, the Continuous Tenses would probably have been introduced to students at an earlier stage. However, in technical language, they are not likely to appear with such great frequency as the Simple Tenses, Active and Passive, already presented. Even oral demonstrations (of experiments, etc.) are more likely to be phrased in the Present Simple than the Present Continuous (see Unit 3).

2 The Present Continuous Active

This tense is introduced in Section A. All the Continuous Tenses in English usually express 'temporariness'. Thus, the Present Continuous will usually indicate an action in progress or of short duration and can be contrasted with the normal use of the Present Simple for 'timeless' statements (see Book 1, Unit 5).

EXAMPLE: The tachometer *indicates* the engine speed (its purpose).
The tachometer *is indicating* the engine speed (at the moment).

At some stage, you may have to warn students that modal verbs (as well as some others, e.g. *know*, *equal*, etc.) are not used in the Continuous Tenses.

GRAMMAR RULES

This tense is formed from the Present Tense of *be*+verb+-*ing*:

I am (*I'm*) go*ing*.
It is (*It's*) work*ing*.

The spelling rules follow other spelling rules previously learnt:

(i) The same verbs that double the final consonant before -*ed* also double it before -*ing*:

travel – travelling – travelled
fit – fitting – fitted
(See Unit 1, Section A, and LANGUAGE NOTE 27.)

(ii) Verbs ending in 'silent' *e* drop the *-e* before *-ing*:

indicate – indicating
produce – producing
(See LANGUAGE NOTE 27.)

(iii) Verbs ending in a consonant+*y*, however, DO NOT change *y* to *i* before *-ing*:

supply – supplying – supplied
vary – varying – varied

Before Listening to the Text

Introduce the Present Continuous with some examples from the text. Contrast its implication of 'temporariness' with the 'timeless' quality of simple statements in the Present Simple.

EXAMPLES: The tachometer indicates engine speed.
The tachometer is indicating . . . . rpm.

The speedometer indicates the road speed.
The speedometer is indicating . . . . k/h.

The ammeter indicates the condition of the battery.
The ammeter is indicating . . . . A.
The battery is charging/discharging.

Further examples can be made up for *thermometer*, *fuel* or *pressure gauge*, etc. With a few simple classroom aids, a visual presentation could be made:

EXAMPLES: Wood burns.
This match is burning.

A clock (*or* watch) indicates the time.
This clock is indicating . . . .
(etc.)

Include some negative sentences.

EXAMPLES: Water boils at 100°C.
This water is not boiling.

Hot metal expands.
This metal is not expanding.
(etc.)

And finish with:

> You are learning English; you are not learning . . . .
> You are listening; you are not reading.
> I am talking; I am not writing.
> (etc.)

The Listening Text

Machines usually require instruments for their control. The instruments provide information about the condition of the machine. For *instance*, there's usually a temperature gauge in a car. If the engine becomes too hot, it'll become damaged. When the temperature gauge *indicates* a very high engine temperature, the engine must be switched off immediately.

Some cars have another instrument called the *tachometer*. Tachometers measure the speed of the engine. Engine speed's measured in rpm. As the engine turns, it also causes the fan to turn. The fan's connected to the alternator by the fan belt. The alternator generates an electric current as it turns. When the car's in use, the alternator recharges the battery and supplies current for the spark plugs, the headlights, etc.

There'll also be an ammeter in some cars. This instrument measures electric current in amperes. In a car, the ammeter *indicates* the flow of current to and from the battery. The scale on the ammeter has positive and negative values. When the needle's *pointing* to a negative value, the battery's discharging. When the needle's pointing to a positive value, the battery's charging. If the engine's turning but the ammeter needle's pointing to zero, then the battery's fully charged.

Exercises

Exercise 1 (Written/Reading)

KEY

1 True.
2 False. . . . many cars have a tachometer.
3 True.
4 False. . . . the speedometer will always indicate zero.
5 True.
6 False. . . . the battery is now recharging.
7 False. . . . the battery will soon be completely discharged.
8 False. . . . the alternator recharges the battery.
9 True.
10 True (fig. 8d).

Exercise 2 (Written)

KEY

1 *Road Speed*: 80 k/h. *Engine Speed*: 4000 rpm.
Alternator: to the battery and the engine.
Ammeter and Battery: +5 A . . . . battery charging.

2 *Road Speed*: 100 k/h. *Engine Speed*: 4500 rpm.
Alternator: to the engine (battery fully charged).
Ammeter and Battery: zero . . . . battery fully charged.

3 *Road Speed*: zero. *Engine Speed*: zero.
Alternator: no current.
Ammeter and Battery: zero . . . . battery not charging or discharging.

4 The engine is not turning. The driver is pushing the car. The car is moving (slowly up a slope).

5 The engine is overheating. The water is boiling. Steam is escaping (from the radiator).

Exercise 3 (Oral)

This exercise reinforces the contrast between the usage in English of the Present Simple and the Present Continuous.

New Vocabulary: *a passenger* *a trailer* *a window*
clean *pull*

Exercise 4 (Reading/Vocabulary)

New Vocabulary: *alternating current* *direct current*
after *a rectifier*

KEY

1 supply	2 soon	3 generate	4 charge
5 either	6 still	7 require	8 because
9 instance	10 efficient	11 although	12 convert

SECTION B: The Multimeter

Teaching Points

1 The Continuous Tenses (cont.)

The Present Continuous Passive

The Passive Continuous, as used here, is similar in usage to the Active form. It can be contrasted with the Simple Passive in the same way as the two Active tenses (Continuous and Simple) were contrasted in Section A.

EXAMPLE: The multimeter *is used* to measure . . . . (its purpose).
The multimeter *is being used* to measure . . . . (at the moment).

Point out to students that, in the Passive Continuous, the lexical verb keeps its *-ed* ending and is preceded by the Continuous form of *be* (*is/are being*).

NOTE: The Passive Continuous, of course, involves irregular verb forms. There is only one irregular verb form in Section B, involving the verb *hold* (*held*), in Exercise 5.

2 Time Subordinators

Before, *after* and *when* are practised in Exercise 7 as subordinators. However, they are practised here in Participial Clauses (i.e. clauses without a finite verb form).

EXAMPLES: Do not hold the metal part *when checking* mains voltage.
When using a multimeter, these two points should be remembered.

NOTE: In speech (and sometimes in writing), clauses of this type are often replaced by the Present Simple:

When using a multimeter. . . .
When you use a multimeter. . . .
(See Exercise 11 in Section C.)

3 Electrical Terminology

A *multimeter* is often referred to as an *AVO meter*.
Ampere (*amperes*) is often shortened to *amp* (*amps*).
The *ohm* is the practical unit of measurement for *electrical resistance*, although, like *volt*, it is not strictly speaking an SI unit and therefore not included in Appendix Two. The ohm symbol is the Greek letter omega (Ω).

a.c. (*alternating current*) and *d.c.* (*direct current*) are the two different forms of electrical supply. Mains electricity is always a.c. whereas small workshop generators, etc. are likely to supply d.c.

The Listening Text

Look at **figure** 8e. It shows a meter. A meter's a device for measuring units. This one's used for measuring electrical units. So its main use, or *function*, is for checking electrical equipment. It can measure voltage or amperage. It can also measure electrical resistance. That's why it's called a *multimeter* – because it has several different functions.

These meters are sometimes called AVO meters – A for amps, V for volts and O for *ohms*. (*Write the last name and its symbol on the board.*) An ohm is the unit of measurement for electrical resistance.

When you use a multimeter, you must first *select* the correct function. (*Repeat in Passive*, *i.e.* When a multimeter *is being used*, the correct function must first be selected. The selector switch on the front must be turned to the correct function.)

If you are measuring *a.c.* voltage, turn the selector to Va.c. (*Repeat*: 'If a.c. voltage is being measured. . . .') If *d.c.* is being measured, turn the selector toVd.c. (*Repeat for amperage* (*A*) *and ohms* (Ω).) Under each function, there are several different ranges of values. *Remember* to select the correct range of values before you use the meter.

For example, if you are measuring mains supply voltage, make sure the selector switch isn't on a low voltage range. If the switch is on the 0–12 volts range when mains voltage is being checked, the meter'll probably be damaged.

Exercises

Exercise 5 (Written/Reading)

Sentences 1–8 can be used orally first as a reading comprehension check if so desired.

KEY (both paragraphs)

The multimeter is being *used*. . . .
The probes are being *held*. . . .
The function selector switch *is turned*. . . .
The needle *is indicating*. . . .

Exercise 6 (Oral/Written)

New Vocabulary: *a soldering iron*

KEY

1 The soldering iron is being connected to the mains supply.
2 The wires are being joined by hand.
3 The joint is being heated with the soldering iron.
4 The solder is being melted with the soldering iron.
5 The joint is being covered with liquid solder.
6 The soldering iron is being replaced on its stand.
7 The plug is being removed from the socket.

Exercise 7 (Oral)

KEY

1 when checking 2 when measuring 3 before connecting
4 after connecting 5 before using 6 after using
7 when testing 8 when checking

Exercise 8 (Reading/Vocabulary)

KEY

1 function 2 insulated 3 terminal 4 points 5 connecting
6 turn 7 than 8 damage 9 instance 10 current 11 direct
12 red 13 or else 14 may 15 being 16 disconnected
17 provides

SECTION C: Battery Chargers

Teaching Points

1 Continuous Tenses (cont.)

The Past Continuous Tenses (Active and Passive)
These indicate 'temporariness' in the same way as the two Present Continuous Tenses. However, unlike the Present Tenses, there is generally an adverb in any clause containing a Past Continuous Tense to show the 'duration' of the activity or event described.

EXAMPLES: *During the first hour*, the needle *was indicating* 5 A.
The technician *was checking* the radio *for 50 minutes.*

However, this is not always the case. The time adverbial may indicate a 'point' in time rather than a 'duration'.

EXAMPLE: *At 8.30*, he *was* (still) *checking* the radio.

But the implication of such a sentence is that the activity lasted over a particular period *which included* the 'time' mentioned. It is possible to contrast the usage of the Past Continuous with that of the Past Simple from this point of view.

EXAMPLES: Parsons *built* his first turbine in 1884 (a completed event).
Parsons *was building* his first turbine in 1884 (and continued to do so after 1884).

NOTE: This 'durative' use of the Past Continuous leads to practice of *during*, *after*, *for* and *at* as well as the new subordinator *while*.

2 Telling the Time
The use of the Past Continuous presents an opportunity to introduce and practise telling the time (see Exercise 12 and LANGUAGE NOTE 30). Although the normal oral forms for telling the time are presented, it is worth noting that the 'numerical' way of telling the time (i.e. 8.15 = eight-fifteen, instead of a quarter-past eight) may be sufficient for most occasions. Especially in recent years, with the widespread marketing of digital watches and clocks, which show the time numerically rather than by means of hands on a face, it is becoming quite common to hear the time stated numerically.

The Listening Text

(*This text does* not *include examples of Past Continuous Tenses.*)

Look at the *battery charger* in the diagram. It's a device for recharging car batteries. It's connected to the mains supply on one side and to the battery terminals on the other. Crocodile clips are used to connect it to the battery terminals.

A battery produces a d.c. current, but the mains supply is always a.c. Moreover, most batteries are only twelve-volt, but mains voltage may be either 110 V or 220 V. So a battery charger has two main functions – to *rectify* a.c. to d.c., and to transform mains voltage to the battery voltage.

When using a battery charger, there are several points to remember. Although most batteries are 12 V nowadays, there are still some 6 V batteries in use. You can see there's a voltage selector on the front of the charger. Check the battery voltage and put the voltage selector in the correct position. Then connect the crocodile clips to the battery terminals. Always connect the clips before you plug the charger into the mains. Also make sure the positive lead from the charger goes to the positive terminal of the battery.

There's an ammeter on the front of the charger inside a square window. It'll indicate the *charging rate while* the battery's charging. It works in the same way as an ammeter in a car. The more discharged the battery is, the faster it'll recharge. When the needle finally falls to zero, the battery's fully charged again.

After Listening to the Text

Telling the time (in the numerical way) was practised with students in Book 1, Unit 2, but this is a good opportunity to revise it with the aid of a clock face – times written on the board, or questions about the school timetable, etc. An introduction to the more traditional oral phrases set out in LANGUAGE NOTE 30 may be postponed until Exercise 12 is being dealt with, if so desired.

Exercises

Exercise 9 (Written/Reading)

In this final Section of Book 2, we have included here some questions which are considerably more open-ended, and require more ambitious use of English in answering than anything attempted so far. Oral discussion of the main point(s) in each answer will be necessary before writing commences.

KEY (main points)

1 No – battery charger required – to rectify a.c. to d.c. – to reduce voltage.
2 By electrical fault – electrical items in use when the engine is not in use (headlights, wipers, radio, etc.).
3 Positive clip on positive terminal – clips connected before mains supply – select correct voltage – secure connections, etc.
4 No – maximum rate in first hour – examples from text.
5 Because battery 'gases' – hydrogen produced – hydrogen inflammable.

Exercise 10 (Oral/Written)

It is possible to state in English that:

The battery *was charging*.

or The battery *was being charged*.

without change of meaning. You may wish to advocate the use of the Passive Tense in this exercise so that each answer will contain one Active and one Passive verb.

NOTE: If not already introduced, the use of *at* with 'points' of time in English must be presented to students for this exercise.

Exercise 11 (Oral/Vocabulary)

KEY

1 making . . . . plugs 2 to . . . . connecting 3 crocodile . . . . check
4 switching . . . . correct 5 near . . . . charging 6 disconnecting
7 replace . . . . battery 8 make sure

Exercise 12 (Oral/Vocabulary)

New Vocabulary: *a (rest) period* *a tea/lunch break* *a time sheet*
a technician *repair* *sign* *approve*
an amplifier *a transistor* *a tape recorder*

First let your students study the time sheet and ask any questions they may have. Then introduce and practise time telling (see LANGUAGE NOTE 30). Finally, ask questions, using the patterns set out in the Students' Book, before allowing students to begin questioning each other using the same patterns.

List of Language Items in Each Unit

Unit One

'-ed' words (adjectives)
'un-' (prefix)
Specialized vocabulary for fitters

Unit Two

Past Simple Tense (regular and irregular verbs)
Possessive ''s'
'And so'
'How'+verb questions
'He' and 'his'
'What did . . . . do?'

Unit Three

Present Simple Tense (*I/you/we*) for demonstrations
Simple Passive Tenses (Present and Past)
Position of adverbs with the Passive
Quantities ('a piece of', etc.)
'Let/Allow'

Unit Four

The Passive with 'by'+agent
Irregular verbs (Passive forms)
'-er/-or' (suffix)
'-ize' and '-ite' (suffixes)

Unit Five

Subordinators (*if*, *as*, *although*, *unless*, *when*)
Days of the week
Months of the year
'Any'

Unit Six

Modal verbs (*will*, *may*, *can*)
Complex noun phrases
Complex verb phrases
'Be' (infinitive)
'Once/twice a' (*every*) . . . .

Unit Seven

Modal verbs (cont.) (*must*, *need*, *not*, *should*)
'In order to, or else' and 'so that'
'-ing' words (nouns and adjectives)
Sign languages (warnings, etc.)
Parts of the body and names of clothes

Unit Eight

The Continuous Tenses (Active/Passive and Present/Past)
'Before', 'after', 'while' (subordinators)
Telling the time

[illegible] of Language Items in [illegible]

[illegible] words [illegible]

[illegible]

[illegible]

[illegible] (regular verbs)

[illegible]

[illegible]

[illegible]

Unit Three [illegible]

[illegible] and [illegible]

[illegible]

[illegible]

[illegible]

[illegible] with [illegible]

[illegible]

[illegible]

[illegible]

[illegible]

[illegible]

[illegible]

[illegible]

[illegible]

[illegible]

[illegible]

[illegible]

Unit Seven [illegible]

[illegible]

[illegible]

[illegible]

[illegible] and [illegible]

[illegible]

[illegible]

[illegible]

[illegible]